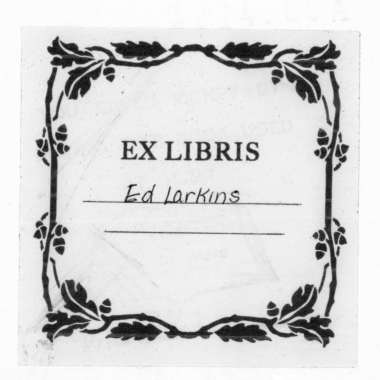

EX LIBRIS

Ed Larkins

Probabilistic Methods
of Signal and
System Analysis

Holt, Rinehart and Winston Series
in Electrical Engineering, Electronics, and Systems

Other Books in the Series:

Shu-Park Chan, Introductory Topological Analysis of Electrical Networks
George R. Cooper and Clare D. McGillem, Methods of Signal and System Analysis
Woodrow W. Everett, Arlon T. Adams, and José Perini, et al., Topics in Intersystem Electromagnetic Compatibility
Mohammed S. Ghausi and John J. Kelly, Introduction to Distributed-Parameter Network: With Application to Integrated Circuits
Roger H. Holmes, Physical Principles of Solid State Devices
Benjamin J. Leon, Lumped Systems
Benjamin J. Leon and Paul A. Wintz, Basic Linear Networks for Electrical and Electronics Engineers
Clare D. McGillem and George R. Cooper, Continuous and Discrete Signal and System Analysis
Richard Saeks, Generalized Networks
Samuel Seeley, Electronic Circuits
Amnon Yariv, Introduction to Optical Electronics

Probabilistic Methods of Signal and System Analysis

GEORGE R. COOPER

Professor of Electrical Engineering

Purdue University
Lafayette, Indiana

CLARE D. McGILLEM

Professor of Electrical Engineering

Purdue University
Lafayette, Indiana

HOLT, RINEHART AND WINSTON, INC.

New York   Chicago   San Francisco   Atlanta
Dallas   Montreal   Toronto   London   Sydney

Library of Congress Catalog Card Number: 73-136170
ISBN: 0-03-084291-3
Printed in the United States of America
1 2 3 4   038   1 2 3 4 5 6 7 8 9

Preface

This book presents an introduction to probability theory, random processes, and the analysis of systems with random inputs. It is written at a level that is suitable for junior and senior engineering students and presumes that the student is familiar with conventional methods of system analysis such as convolution and transform techniques. However, it may also serve graduate students as a concise review of material that they previously encountered in widely scattered sources.

Since this is an engineering text, the treatment is heuristic rather than rigorous, and the student will find many examples of the application of these concepts to engineering problems. However, it is not completely devoid of the mathematical subtleties, and considerable attention has been devoted to pointing out some of the difficulties that make a more advanced study of the subject essential if one is to master it. The authors believe that the educational process is best served by repeated exposure to difficult subject matter; this text is intended to be the first exposure to probability and random processes, but hopefully, will not be the last. Thus, the book is not comprehensive, but deals selectively with those topics that the authors have found most useful in the solution of engineering problems.

A brief discussion of some of the significant features of this book will help set the stage for a discussion of the various ways it can be used.

Elementary concepts of discrete probability are introduced in Chapter 1; first from the intuitive standpoint of the relative-frequency approach and then from the more rigorous standpoint of axiomatic probability. Simple examples illustrate all these concepts and are more meaningful to engineers than are the traditional examples of selecting red and white balls from urns.

The concept of a random variable is introduced in Chapter 2 along with the ideas of probability distribution and density functions, mean values, and conditional probability. A significant feature of this chapter is a rather extensive discussion of many different probability density functions and the physical situations in which they may occur. Chapter 3 extends the random variable concept to situations involving two or more random variables and introduces the concepts of statistical independence and correlation.

A general discussion of random processes and their classification is given in Chapter 4. The emphasis here is on selecting probability models that are useful in solving engineering problems. Accordingly, a great deal of attention is devoted to the physical significance of the various process classifications, with no attempt at mathematical rigor. A unique feature of this chapter, which is continued in subsequent chapters, is an introduction to the practical problem of estimating the mean of a random process from an observed sample function.

Properties and applications of autocorrelation and cross-correlation functions are discussed in Chapter 5. Many examples are presented in an attempt to develop some insight into the nature of correlation functions. The important problem of estimating autocorrelation functions is discussed in some detail.

Chapter 6 turns to a frequency-domain representation of random processes by introducing the concept of spectral density. Unlike most texts, which simply define spectral density as a Fourier transform of the correlation function, a more fundamental approach is adopted here in order to bring out the physical significance of the concept. This chapter is the most difficult one in the book, but the authors believe the material should be presented in this way. Instructors who wish to by-pass some of the more fundamental problems may omit Section 6-2 and bridge the gap by defining spectral density simply as the Fourier transform of the correlation function.

Chapter 7 utilizes the concepts of correlation functions and spectral density to analyze the response of linear systems to random inputs. In a sense, this chapter is a culmination of all that preceded it, and is particularly significant to engineers who must use these concepts. Hence, it contains a great many examples that are relevant to engineering problems and emphasizes the need for mathematical models that both are realistic and manageable.

Chapter 8 extends the concepts of systems analysis to consider systems that are optimum in some sense. Both the classical matched filter for known signals and the Wiener filter for random signals are considered from an elementary standpoint.

In a more general vein, each chapter contains references that the student can use to extend his knowledge. There is also a wide selection of problems at the end of each chapter, and, at the end of the book, a number of Appendixes that he will find useful in solving these problems, along with a set of selected answers,

As an additional aid to learning and using the concepts and methods discussed in this book, there are exercises at the end of many of the sections. The student should consider these exercises as part of the reading assignments and should make every effort to solve each one before going on to the next section. Answers are provided in order that he may know when his efforts have been successful. *It should be noted, however, that the answers to each exercise may not be listed in the same order as the question.* This is intended to provide an additional challenge to the student. The presence of these exercises should substantially reduce the number of additional problems that need to be assigned by the instructor.

The material in this text has been used in a one-semester, three-credit course offered in the first semester of the junior year. Not all sections of the text are used in this course but at least 90% of it is covered in reasonable detail. The sections usually omitted include 3-6, 4-6, 5-4, 5-9, 6-9, and 8-6; but other choices may be made at the discretion of the instructor. There are, of course, many other ways in which the text material could be utilized. For example, a one-semester course with a more relaxed pace could be given by omitting all of Chapter 8 in addition to the sections noted above. For those schools on the quarter-system, the material noted above could be covered in a four-credit hour course. If a three-credit hour course were desired, in addition to the omissions noted above, it is suggested that Sections 1-4, 1-5, 1-6, 1-8, 2-6, 2-7, 3-5, 6-2, 6-8, 6-10, 7-9, and all of Chapter 8 can be omitted if the instructor supplies a few explanatory words to bridge gaps. Obviously, there are also many other possibilities that are open to the experienced instructor.

It is a pleasure for the authors to acknowledge the very substantial aid and encouragement that they have received from their colleagues and students. A complete list is too lengthy to include here, but it is appropriate to mention the valuable suggestions and comments received from Professors J. Y. S. Luh and P. A. Wintz of Purdue University.

The careful and perceptive readings of the preliminary manuscript by Professor J. E. Kemmerly of the California State College at Fullerton and Professor James L. Massey of Notre Dame University are gratefully acknowledged. Their many suggestions have greatly improved the final version. A special note of thanks is due to Mr. Lewis A. Thurman, of Purdue University, for his diligent efforts in proofreading, working problems, and making suggestions. Last, but not least, we acknowledge the contributions made by hundreds of students who used and criticized the earliest versions of this material.

July 1971 GEORGE R. COOPER
CLARE D. McGILLEM

Contents

Preface   **v**

Chapter 1   **Introduction to Probability**   **1**
   1-1   Engineering Applications of Probability   1
   1-2   Definitions of Probability   5
   1-3   The Relative-Frequency Approach   6
   1-4   Elementary Set Theory   12
   1-5   The Axiomatic Approach   17
   1-6   Conditional Probability   19
   1-7   Independence   24
   1-8   Combined Experiments   25
   1-9   Bernoulli Trials   27
       Problems   30
       References   33

Chapter 2   **Random Variables**   **35**
   2-1   Concept of a Random Variable   35
   2-2   Distribution Functions   37
   2-3   Density Functions   39
   2-4   Mean Values and Moments   43
   2-5   The Gaussian Random Variable   46

2-6   Density Functions Related to Gaussian   49
2-7   Other Probability Density Functions   55
2-8   Conditional Distribution and Density
        Functions   60
2-9   Examples and Applications   64
        Problems   69
        References   73

Chapter 3   Several Random Variables   74
3-1   Two Random Variables   74
3-2   Conditional Probability—Revisited   78
3-3   Statistical Independence   82
3-4   Correlation Between Random Variables   83
3-5   Density Function of the Sum of Two Random
        Variables   85
3-6   The Characteristic Function   88
        Problems   92
        References   94

Chapter 4   Random Processes   95
4-1   Introduction   95
4-2   Continuous and Discrete Random
        Processes   96
4-3   Deterministic and Nondeterministic
        Random Processes   97
4-4   Stationary and Nonstationary Random
        Processes   98
4-5   Ergodic and Nonergodic Random
        Processes   100
4-6   Measurement of Process Parameters   100
        Problems   104
        References   106

Chapter 5   Correlation Functions   107
5-1   Introduction   107
5-2   Example: Autocorrelation Function of a
        Binary Process   110
5-3   Properties of Autocorrelation Functions   112
5-4   Measurement of Autocorrelation
        Functions   115
5-5   Examples of Autocorrelation Functions   117
5-6   Cross-correlation Functions   119
5-7   Properties of Cross-correlation Functions   120

5-8   Examples and Applications of Cross-correlation
       Functions   122
5-9   Correlation Matrices for Sampled
       Functions   124
       Problems   127
       References   130

Chapter 6   Spectral Density   131
6-1   Introduction   131
6-2   Relation of Spectral Density to the
       Fourier Transform   133
6-3   Properties of Spectral Density   136
6-4   Spectral Density and the Complex
       Frequency Plane   141
6-5   Mean-Square Values from Spectral
       Density   142
6-6   Relation of Spectral Density to the
       Autocorrelation Function   146
6-7   White Noise   150
6-8   Cross-spectral Density   152
6-9   Measurement of Spectral Density   153
6-10   Examples and Applications of Spectral
       Density   158
       Problems   163
       References   166

Chapter 7   Response of Linear Systems to Random Inputs   167
7-1   Introduction   167
7-2   Analysis in the Time Domain   168
7-3   Mean and Mean-Square Value of System
       Output   169
7-4   Autocorrelation Function of System
       Output   172
7-5   Cross-correlation Between Input and
       Output   176
7-6   Example of Time-Domain System
       Analysis   179
7-7   Analysis in the Frequency Domain   183
7-8   Spectral Density at the System Output   183
7-9   Cross-spectral Densities Between Input
       and Output   187
7-10   Examples of Frequency-Domain
       Analysis   187
       Problems   192
       References   195

Chapter 8   **Optimum Linear Systems**   **196**
  8-1   Introduction   196
  8-2   Criteria of Optimality   197
  8-3   Restrictions on the Optimum System   199
  8-4   Optimization by Parameter Adjustment   199
  8-5   Systems That Maximize Signal-to-Noise
           Ratio   205
  8-6   Systems That Minimize Mean-Square
           Error   210
           Problems   215
           References   217

Appendixes   **219**

Appendix A   **Mathematical Tables**   **221**
  A-1   Trigonometric Identities   221
  A-2   Indefinite Integrals   222
  A-3   Definite Integrals   223
  A-4   Fourier Transforms   224
  A-5   One-sided Laplace Transforms   225

Appendix B   **Frequently Encountered Probability Distribution**   **226**
           Discrete Probability Functions   226
           Continuous Distributions   227

Appendix C   **Binomial Coefficients**   **231**

Appendix D   **Normal Probability Distribution Function**   **232**

Appendix E   **Table of Correlation Function—Spectral
           Density Pairs**   **233**

Appendix F   **Contour Integration**   **234**

           Selected Answers   **241**

           Index   **251**

Probabilistic Methods of Signal and System Analysis

1

Introduction to Probability

1-1 Engineering Applications of Probability

Before embarking on a study of elementary probability theory, it is desirable to motivate such a study by considering why probability theory is useful in the solution of engineering problems. This will be done in two different ways. The first is to suggest a viewpoint, or philosophy, concerning probability that emphasizes its universal physical reality rather than treating it as another mathematical discipline which may be useful occasionally. The second is to note some of the many different types of situations that arise in normal engineering practice in which the use of probability concepts is indispensible.

A characteristic feature of probability theory is that it concerns itself with situations that involve uncertainty in some form. The popular conception of this relates probability to such activities as tossing dice, drawing cards, and spinning roulette wheels. Because the rules of probability are not widely known, and because such situations can become quite complex, the prevalent attitude is that probability theory is a mysterious and esoteric branch of mathematics that is accessible only to trained mathematicians and is of only limited value in the real world. Since probability theory does deal with uncertainty, another prevalent attitude is that a probabilistic treatment of physical problems is an inferior substitute for a more desirable exact analysis and is forced on the analyst by a lack of complete information. *Both of these attitudes are false.*

Regarding the alleged difficulty of probability theory, it is doubtful if there is any other branch of mathematics or analysis which is so completely based on such a small number of basic concepts that are so easily understood. Subsequent discussion will reveal that the major body of probability theory can be deduced from only three axioms that are almost self-evident. Once these axioms and their applications are understood, the remaining concepts follow in a logical manner.

The attitude that regards probability theory as a substitute for exact analysis stems from the current educational practice of presenting physical laws as deterministic, immutable, and strictly true under all circumstances. Thus, a law that describes the response of a dynamical system is supposed to predict that response exactly if the system excitation is known exactly. For example, Ohm's law

$$v(t) = Ri(t)$$

is assumed to be exactly true at every instant of time, and on a macroscopic basis this assumption may be well justified. On a microscopic basis, however, this assumption is patently false—a fact that is immediately obvious to anyone who has tried to connect a large resistor to the input of a high-gain amplifier and listened to the resulting noise.

In the light of modern physics and our emerging knowledge of the nature of matter, the viewpoint that natural laws are deterministic and exact is untenable. They are at best a representation of the average behavior of nature. In many important cases this average behavior is close enough to that actually observed so that the deviations are unimportant. In such cases the deterministic laws are extremely valuable because they make it possible to predict system behavior with a minimum of effort. In other equally important cases the random deviations may be significant—perhaps even more significant than the deterministic response. For these cases, analytic methods derived from the concepts of probability are essential.

From the above discussion it should be clear that the so-called exact solution is not exact at all, but in fact represents an idealized special case which actually never arises in nature. The probabilistic approach, on the other hand, far from being a poor substitute for exactness, is actually the method which most nearly represents physical reality. Furthermore, it includes the deterministic result as a special case.

It is now appropriate to discuss the types of situations in which probability concepts arise in engineering. The examples presented here emphasize situations that arise in systems studies; but they do serve to illustrate the essential point that engineering applications of probability tend to be the rule rather than the exception.

Random input signals. In order for a physical system to perform a useful task, it is usually necessary that some sort of forcing function (the input signal) be applied to it. Input signals that have simple mathematical representations are convenient for pedagogical purposes or for certain types of system analysis, but they seldom arise in actual applications. Instead, the input signal is more likely to involve a certain amount of uncertainty and unpredictability that justifies treating it as a *random* signal. There are many examples of this: speech and music signals that serve as inputs to communication systems; random digits applied to a computer; random command signals applied to an aircraft flight control system; random signals derived from measuring some characteristic of a manufactured product, and used as inputs to a process control system; steering wheel movements in an automobile power-steering system; the sequence in which the call and operating buttons of an elevator are pushed; the number of vehicles passing various checkpoints in a traffic control system; outside and inside temperature fluctuations as inputs to a building heating and airconditioning system; and many others.

Random disturbances. Many systems have unwanted disturbances applied to their input or output in addition to the desired signals. Such disturbances are almost always random in nature and call for the use of probabilistic methods even if the desired signal does not. A few specific cases serve to illustrate several different types of disturbances. If, for a first example, the output of a high-gain amplifier is connected to a loudspeaker, one frequently hears a variety of snaps, crackles, and pops. This random noise arises from thermal motion of the conduction electrons in the amplifier input circuit or from random variations in the number of electrons (or holes) passing through the tubes and transistors. It is obvious that one cannot hope to calculate the value of this noise at every instant of time since this value represents the combined effects of literally billions of individual moving charges. It is possible, however, to calculate the average power of this noise, its frequency spectrum, and even the probability of observing a noise value larger than some specified value. As a practical matter, these quantities are more important in determining the quality of the amplifier than is a knowledge of the instantaneous waveforms.

As a second example, consider a radio or television receiver. In addition to noise generated within the receiver by the mechanisms noted, there is random noise arriving at the antenna. This results from distant electrical storms, man-made disturbances, radiation from space, or thermal radiation from surrounding objects. Hence, even if perfect receivers and amplifiers were available, the received signal would be combined with random noise. Again, the calculation of such quantities as average power and frequency spectrum may be more significant than the determination of instantaneous value.

A different type of system is illustrated by a large radar antenna, which may be pointed in any direction by means of an automatic control system. The wind blowing on the antenna produces random forces that must be compensated for by the control system. Since the compensation is never perfect, there is always some random fluctuation in the antenna direction; it is important to be able to calculate the effective value of this fluctuation.

A still different situation is illustrated by an airplane flying in turbulent air, a ship sailing in stormy seas, or an army truck traveling over rough terrain. In all these cases random disturbing forces, acting on complex mechanical systems, interfere with the proper control or guidance of the system. It is important to be able to determine how the system responds to these random input signals.

Random system characteristics. The system itself may have characteristics that are unknown and that vary in a random fashion from time to time. Some typical examples are: aircraft in which the load (that is, the number of passengers or the weight of the cargo) varies from flight to flight; troposcatter communication systems in which the path attenuation varies radically from moment to moment; an electric power system in which the load (that is, the amount of energy being used) fluctuates randomly; and a telephone system in which the number of users changes from instant to instant.

System reliability. All systems are composed of many individual elements, and one or more of these elements may fail, thus causing the entire system, or part of the system, to fail. The times at which such failures will occur is unknown, but it is often possible to determine the probability of failure for the individual elements and from these to determine the "mean time to failure" for the system. Such reliability studies are deeply involved with probability and are extremely important in engineering design. As systems become more complex, more costly, and contain larger numbers of elements, the problems of reliability become more difficult and take on added significance.

Quality control. An important method of improving system reliability is to improve the quality of the individual elements, and this can often be done by an inspection process. As it may be too costly to inspect every element after every step during its manufacture, it is necessary to develop rules for inspecting elements selected at random. These rules are based on probabilistic concepts and serve the valuable purpose of maintaining the quality of the product with the least expense.

Information theory. A major objective of information theory is to provide a quantitative measure for the information content of messages such as

printed pages, speech, pictures, graphical data, numerical data, or physical observations of temperature, distance, velocity, radiation intensity,
and rainfall.   This quantitative measure is necessary in order to be able
to provide communication channels that are both adequate and efficient
for conveying this information from one place to another.   Since such
messages and observations are almost invariably unknown in advance
and random in nature, they can be described only in terms of probability.
Hence, the appropriate information measure is a probabilistic one.
Furthermore, the communication channels are subject to random disturbances (noise) that limit their ability to convey information, and
again a probabilistic description is required.

It should be clear from the above partial listing that almost any
engineering endeavor involves a degree of uncertainty or randomness
that makes the use of probabilistic concepts an essential tool for the
present-day engineer.   In the case of system analysis, it is necessary
to have some description of random signals and disturbances.   There
are two general methods of describing random signals mathematically.
The first, and most basic, is a probabilistic description in which the
random quantity is characterized by a probability model.   This method
is discussed later in this chapter.

The probabilistic description of random signals cannot be used directly
in system analysis since it tells very little about how the random signal
varies with time or what its frequency spectrum is.   It does, however,
lead to the statistical description of random signals, which is useful in
system analysis.   In this case the random signal is characterized by a
statistical model, which consists of an appropriate set of average values
such as the mean, variance, correlation function, spectral density, and
others.   These average values represent a less precise description of the
random signal than that offered by the probability model, but they are
more useful for system analysis because they can be computed by using
straightforward and relatively simple methods.   Some of the statistical
averages will be discussed in subsequent chapters.

1-2   Definitions of Probability

One of the most serious stumbling blocks in the study of elementary
probability is that of arriving at a satisfactory definition of the term
"probability."   There are, in fact, four or five different definitions for
probability that have been proposed and used with varying degrees of
success.   They all suffer from deficiencies in concept or application.
Ironically, the most successful "definition" leaves the term probability
undefined.

Of the various approaches to probability, the two that appear to
be most useful are the *relative-frequency* approach and the *axiomatic*

approach.   The relative-frequency approach is useful because it attempts to attach some physical significance to the concept of probability and thereby makes it possible to relate probabilistic concepts to the real world.   Hence the application of probability to engineering problems is almost always accomplished by invoking the concepts of relative frequency, even when the engineer may not be conscious that he is doing so.

The limitation of the relative-frequency approach is the difficulty of using it to deduce the appropriate mathematical structure for situations that are too complicated to be analyzed readily by physical reasoning. This is not to imply that this approach cannot be used in such situations, for it can, but it does suggest that there may be a much easier way to deal with these cases.   The easier way turns out to be the axiomatic approach.

The axiomatic approach treats the probability of an event as a number that satisfies certain postulates but is otherwise undefined.   Whether or not this number relates to anything in the real world is of no concern in developing the mathematical structure that evolves from these postulates.   Engineers may object to this approach as being too artificial and too removed from reality, but they should remember that the whole body of circuit theory was developed in essentially the same way.   In the case of circuit theory the basic postulates are Kirchhoff's laws and the conservation of energy.   The same mathematical structure emerges regardless of what physical quantities are identified with the abstract symbols—or even if *no* physical quantities are associated with them. It is the task of the engineer to relate this mathematical structure to the real world in a way that is admittedly not exact, but that leads to useful solutions to real problems.

From the above discussion it appears that the most useful approach to probability for engineers is a two-pronged one, in which the relative-frequency concept is employed in order to relate simple results to physical reality, and the axiomatic approach is employed to develop the appropriate mathematics for more complicated situations.   It is this philosophy that will be presented here.

1-3   The Relative-Frequency Approach

As its name implies, the relative-frequency approach to probability is closely linked to the frequency of occurrence of some particular event. The term *event* is used for one of the most basic concepts of probability theory.   An event is something that may or may not happen.   For example, if a coin is tossed the result may be a head or a tail and each of these is an event.

In order to examine this concept more precisely, however, it is necessary to introduce the idea of an *experiment* and the *outcomes* of that experiment.   Although there is a precise mathematical definition for an experiment, a better understanding may be gained by listing some examples of well-defined experiments and their possible outcomes.   This is done in Table 1-1.   It should be noted, however, that the possible

Table 1-1 POSSIBLE EXPERIMENTS AND THEIR OUTCOMES

EXPERIMENT	POSSIBLE OUTCOMES
Flipping a coin	Heads (H), tails (T)
Throwing a die	1, 2, 3, 4, 5, 6
Drawing a card	The 52 possible cards
Observing a voltage	Greater than zero, less than zero
Observing a voltage	Greater than V, less than V
Observing a voltage	Between V_1 and V_2, not between V_1 and V_2

outcomes often may be defined in several different ways depending upon the wishes of the experimenter.   The initial discussion is concerned with only a single performance of a well-defined experiment.   This single performance is referred to as a *trial*.

In all of the experiments listed in Table 1-1 the number of outcomes is finite.   This situation pertains to the special case of *discrete probability*, which will provide the starting point for the present discussion.   It is worth noting, however, that in many cases an infinite number of outcomes are possible.   For example, in observing a voltage, there may be a continuous range of possible values that might be observed.   Our major interest will be with situations that involve *continuous probability*.   Since all of the terminology and methods of the discrete case carry over to the continuous case, it is reasonable to consider the discrete case first.

An event is one or more outcomes that are of particular interest to the experimenter.   For example, in throwing a die, each of the outcomes 1, 2, . . . , 6 may be considered as an event.   Alternatively, we may choose to define events as obtaining an even number, or obtaining an odd number, or obtaining a number greater than three, or any of a large number of other possible combinations of outcomes.

It is also possible to conduct more complicated experiments with more complicated sets of events.   The experiment may consist of tossing ten coins, and it is apparent in this case that there are many different possible outcomes, each of which may be an event.   Another situation, which has more of an engineering flavor, is that of a telephone system having 10,000 telephones connected to it.   At any given time, a possible event is that 2000 of these telephones are in use.   Obviously, there are a great many other possible events.

If the outcome of an experiment is uncertain before the experiment is performed, the possible outcomes are random events.   To each of these events it is possible to assign a number, called the probability of that event, and this number is a measure of how likely that event is.   Usually, these numbers are assumed, the assumed values being based on our intuition about the experiment. For example, if we toss a coin, we would expect that the possible outcomes of *heads* and *tails* would be equally likely. Therefore, we would assume the probabilities of these two events to be the same.

It was noted earlier that probability could be related to the relative frequency with which an event occurred. Thus, in tossing a coin, we expect a head to occur half the time and a tail half the time. Hence, we assign a probability of one half to each of these events. More generally, if an experiment is performed N times and we expect event A to occur N_A times, then we assume the probability of event A, Pr (A), to be

$$\Pr (A) = \frac{N_A}{N} \tag{1-1}$$

Note that N_A is not the actual number of times that event A occurs in N trials, but only the number we assume occur on the basis of our intuition about the experiment. We could evaluate the validity of our assumption by performing the experiment and observing the number of times, N'_A, the event A actually occurs. If our assumption is correct, then the ratio N'_A/N approaches $\Pr(A)$ as N becomes large. But making an evaluation of this sort is a problem of statistics rather than probability and does not concern us here.

The relative-frequency approach to probability does make evident a number of important properties of probability. Since both N_A and N are positive, real numbers and N_A must be less than or equal to N; it follows that a probability cannot be less than zero nor greater than unity. Furthermore, suppose the experiment has possible outcomes, $A, B, \ldots, M$; only one of which can occur on any given trial. The possible events are said to be mutually exclusive. If we expect event A to occur N_A times out of N trials, event B to occur N_B times, and so forth, then

$$N_A + N_B + N_C + \cdots + N_M = N$$

or

$$\frac{N_A}{N} + \frac{N_B}{N} + \frac{N_C}{N} + \cdots + \frac{N_M}{N} = 1$$

From the relation given above, it follows that

$$\Pr (A) + \Pr (B) + \Pr (C) + \cdots + \Pr (M) = 1 \tag{1-2}$$

and we can conclude that the sum of the probabilities of all of the mutually exclusive events associated with a given experiment must be unity.

These concepts can be summarized by the following set of statements:

1. $0 \leq \Pr (A) \leq 1$.

2. $\Pr (A) + \Pr (B) + \Pr (C) + \cdots + \Pr (M) = 1$, for a complete set of mutually exclusive events.

3. An impossible event is represented by Pr $(A) = 0$.

4. A certain event is represented by Pr $(A) = 1$.

In order to make some of these ideas more specific, consider the following hypothetical example. Assume that a large bin contains an assortment of resistors of different sizes, which are thoroughly mixed. In particular, let there be 100 resistors having a marked value of 1 Ω, 500 resistors marked 10 Ω, 150 resistors marked 100 Ω, and 250 resistors marked 1000 Ω. Someone reaches into the bin and pulls out one resistor at random. There are now four possible outcomes corresponding to the value of the particular resistor selected. We desire to determine the probability of each of these events. In order to do this, we assume that the probability of each event is proportional to the number of resistors in the bin corresponding to that event. Since there are 1000 resistors in the bin all together, the resulting probabilities are

$$\text{Pr } (1 \ \Omega) = \frac{100}{1000} = 0.1 \qquad \text{Pr } (10 \ \Omega) = \frac{500}{1000} = 0.5$$

$$\text{Pr } (100 \ \Omega) = \frac{150}{1000} = 0.15 \qquad \text{Pr } (1000 \ \Omega) = \frac{250}{1000} = 0.25$$

Note that these probabilities are all positive, less than 1, and do add up to 1.

Many times one is interested in more than one event at a time. If a coin is tossed twice, one may wish to determine the probability that a head will occur on both tosses. Such a probability is referred to as a joint probability. In this particular case, one assumes that all four possible outcomes (HH, HT, TH, and TT) are equally likely and, hence, the probability of each is one quarter. In a more general case the situation is not this simple, so it is necessary to look at a more complicated situation in order to deduce the true nature of joint probability. The notation employed is Pr (A,B) and signifies the probability of the joint occurrence of events A and B.

Consider again the bin of resistors and specify that in addition to having different resistance values, they also have different power ratings. Let the different power ratings be 1 W, 2 W, and 5 W; the number having each rating is indicated in Table 1-2.

Table 1-2 RESISTANCE VALUES AND POWER RATINGS

POWER RATING	RESISTANCE VALUES				
	1 Ω	10 Ω	100 Ω	1000 Ω	TOTALS
1 W	50	300	90	0	440
2 W	50	50	0	100	200
5 W	0	150	60	150	360
Totals	100	500	150	250	1000

Before using this example to illustrate joint probabilities, consider the probability (now referred to as a marginal probability) of selecting a resistor having a given power rating without regard to its resistance value. From the totals given in the right-hand column, it is clear that these probabilities are

$$\text{Pr (1 W)} = \frac{440}{1000} = 0.44 \qquad \text{Pr (2 W)} = \frac{200}{1000} = 0.20$$

$$\text{Pr (5 W)} = \frac{360}{1000} = 0.36$$

We now ask what the joint probability is of selecting a resistor of 10 Ω having a 5-W power rating. Since there are 150 such resistors in the bin, this joint probability is clearly

$$\text{Pr (10 Ω, 5 W)} = \frac{150}{1000} = 0.15$$

The eleven other joint probabilities can be determined in a similar way. Note that some of the joint probabilities are zero [for example, Pr (1 Ω, 5 W) = 0] simply because a particular combination of resistance and power does not exist.

It is necessary at this point to relate the joint probabilities to the marginal probabilities. In the example of tossing a coin two times, the relationship is simply a product. That is,

$$\text{Pr (H,H)} = \text{Pr (H) Pr (H)} = \frac{1}{2} \times \frac{1}{2} = \frac{1}{4}$$

But this relationship is obviously not true for the resistor bin example. Note that

$$\text{Pr (5 W)} = \frac{360}{1000} = 0.36$$

and it was previously shown that

$$\text{Pr (10 Ω)} = 0.5$$

Thus,

$$\text{Pr (10 Ω) Pr (5 W)} = 0.5 \times 0.36 = 0.18 \neq \text{Pr (10 Ω, 5 W)} = 0.15$$

and the joint probability is not the product of the marginal probabilities.

In order to clarify this point, it is necessary to introduce the concept of conditional probability. This is the probability of one event A, given that another event B has occurred; it is designated as $\text{Pr }(A|B)$. In terms of the resistor bin, consider the conditional probability of selecting a 10-Ω resistor when it is already known that the chosen resistor is 5 W. Since there are 360 5-W resistors, and 150 of these are 10 Ω, the required conditional probability is

$$\Pr\ (10\ \Omega|5\ \mathrm{W})\ =\ \frac{150}{360}\ =\ 0.417$$

Now consider the product of this conditional probability and the marginal probability of selecting a 5-W resistor.

$$\Pr\ (10\ \Omega|5\ \mathrm{W})\ \Pr\ (5\ \mathrm{W})\ =\ 0.417\ \times\ 0.36\ =\ 0.15\ =\ \Pr\ (10\ \Omega,\ 5\ \mathrm{W})$$

It is seen that this product is indeed the joint probability.

The same result can also be obtained another way. Consider the conditional probability

$$\Pr\ (5\ \mathrm{W}|10\ \Omega)\ =\ \frac{150}{500}\ =\ 0.30$$

since there are 150 5-W resistors out of the 500 10-Ω resistors. Then form the product

$$\Pr\ (5\ \mathrm{W}|10\ \Omega)\ \Pr\ (10\ \Omega)\ =\ 0.30\ \times\ 0.5\ =\ \Pr\ (10\ \Omega,\ 5\ \mathrm{W}) \qquad \text{(1-3)}$$

Again, the product is the joint probability.

The foregoing ideas concerning joint probability can be summarized in the general equation

$$\Pr\ (A,B)\ =\ \Pr\ (A|B)\ \Pr\ (B)\ =\ \Pr\ (B|A)\ \Pr\ (A) \qquad \text{(1-4)}$$

which indicates that the joint probability of two events can always be expressed as the product of the marginal probability of one event and the conditional probability of the other event given the first event.

We now return to the coin-tossing problem, in which it is indicated that the joint probability can be obtained as the product of two marginal probabilities. Under what conditions will this be true? From equation (1-4) it appears that this can be true if

$$\Pr\ (A|B)\ =\ \Pr\ (A) \qquad \text{and} \qquad \Pr\ (B|A)\ =\ \Pr\ (B)$$

These statements imply that the probability of event A does not depend upon whether or not event B has occurred. This is certainly true in coin tossing, since the outcome of the second toss cannot be influenced in any way by the outcome of the first toss. Such events are said to be statistically independent. More precisely, two random events are statistically independent if and only if

$$\Pr\ (A,B)\ =\ \Pr\ (A)\ \Pr\ (B) \qquad \text{(1-5)}$$

The preceding paragraphs provide a very brief discussion of many of the basic concepts of discrete probability. They have been presented in a heuristic fashion without any attempt to justify them mathematically. Instead, all of the probabilities have been formulated by invoking the concept of relative frequency and in terms of specific numerical examples. It is clear from these examples that it is not difficult to assign reasonable numbers to the probabilities of various events (by employing the relative-frequency approach) when the physical situation is not very involved. It should also be apparent, however, that such an approach might become unmanageable when there are many possible outcomes to any experiment and many different ways of defining events. This is particularly true when one attempts to extend the results for the discrete case to the continuous case. It becomes necessary, therefore, to reconsider all of the above ideas in a more precise manner and to introduce a measure of mathematical rigor that provides a more solid footing for subsequent extensions.

Exercise 1-3

It is known that there are two bad transistors in a group of ten. A transistor is selected from the group and measured and then a second transistor is selected and measured. What is the probability that (a) both are good, (b) the second one is defective, (c) both are defective?

Answer: 1/5, 28/45, 1/45

(Note: In this exercise, and in others throughout the book, the answers are not necessarily given in the same order as the questions.)

1-4 Elementary Set Theory

The more precise formulation mentioned in Section 1-3 is accomplished by putting the ideas introduced in that section into the framework of the axiomatic approach. In order to do this, however, it is first necessary to review some of the elementary concepts of set theory.

A set is a collection of objects known as *elements*. It will be designated as

$$A = \{\alpha_1, \alpha_2, \ldots, \alpha_n\} \tag{1-6}$$

where the set is A and the elements are $\alpha_1, \ldots, \alpha_n$. The set A may consist of the integers from 1 to 6 so that $\alpha_1 = 1$, $\alpha_2 = 2$, $\ldots$, $\alpha_6 = 6$ are the elements. A subset of A is any set all of whose elements are also elements of A. $B = \{1,2,3\}$ is a subset of the set $A = \{1,2,3,4,5,6\}$. The general notation for indicating that B is a subset of A is $B \subset A$. Note that every set is a subset of itself.

All sets of interest in probability theory have elements taken from the largest set called a *space* and designated as S. Hence, all sets will be subsets of the space S. The relation of S and its subsets to probability will become clear shortly, but in the meantime an illustration may be helpful. Suppose that the elements of a space consist of the six faces of a die, and that these faces are designated as 1, 2, $\ldots$, 6. Thus,

$$S = \{1,2,3,4,5,6\}$$

There are many ways in which subsets might be formed, depending upon the number of elements belonging to each subset.   In fact, if one includes the *null set* or *empty set*, which has no elements in it and is denoted by $\emptyset$, there are $64 = 2^6$ subsets and they may be denoted as

$$\emptyset, \{1\}, \ldots, \{6\}, \{1,2\}, \{1,3\}, \ldots, \{5,6\}, \{1,2,3\}, \ldots, S$$

In general, if S contains n elements, then there are 2^n subsets.   The proof of this is left as an exercise for the student.

One of the reasons for using set theory to develop probability concepts is that the important operations are already defined for sets and have simple geometric representations that aid in visualizing and understanding these operations.   The geometric representation is the *Venn diagram* in which the space S is represented by a square and the various sets are represented by closed plane figures.   For example, the Venn diagram shown in Figure 1-1 shows that B is a subset of A and that C

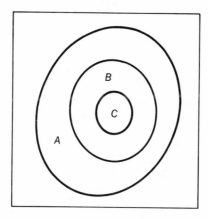

Figure 1-1   Venn diagram for $C \subset B \subset A$.

is a subset of B (and also of A).   The various operations are now defined and represented by Venn diagrams.

Equality.   Set A equals set B *iff* (if and only if) every element of A is an element of B and every element of B is an element of A.   Thus

$$A = B \quad \text{iff} \quad A \subset B \quad \text{and} \quad B \subset A$$

The Venn diagram is obvious and will not be shown.

Sums.   The *sum* or *union* of two sets is a set consisting of all the elements that are elements of A or of B or of both.   This is shown in Figure 1-2. Since the associative law holds, the sum of more than two sets can be written without parentheses.   That is

$$(A + B) + C = A + (B + C) = A + B + C$$

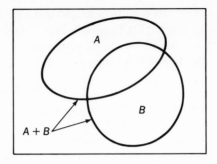

Figure 1-2   The sum of two sets, $A + B$.

The commutative law also holds, so that

$$A + B = B + A$$
$$A + A = A$$
$$A + \varnothing = A$$
$$A + S = S$$
$$A + B = A, \text{ if } B \subset A$$

Products.   The *product* or *intersection* of two sets is the set consisting of all the elements that are common to both sets.   It is designated as AB and is illustrated in Figure 1-3.   A number of results apparent from the

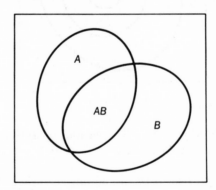

Figure 1-3   The product of two sets, AB.

Venn diagram are

$$AB = BA \qquad \text{(Commutative law)}$$
$$AA = A$$
$$A\varnothing = \varnothing$$
$$AS = A$$
$$AB = B, \text{ if } B \subset A$$

If there are more than two sets involved in the product, the Venn

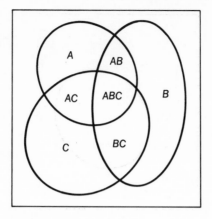

Figure 1-4   Products for three sets.

diagram of Figure 1-4 is appropriate.   From this it is seen that

$$(AB)C = A(BC) = ABC$$
$$A(B + C) = (AB) + (AC) \qquad \text{(Associative law)}$$

Two sets A and B are *mutually exclusive* or *disjoint* if $AB = \varnothing$. Representations of such sets in the Venn diagram do not overlap.

Complement.   The complement of a set A is a set containing all the elements of S that are *not* in A.   It is denoted $\bar{A}$ and is shown in Figure 1-5.   It is clear that

$$\bar{\varnothing} = S$$
$$\bar{S} = \varnothing$$
$$(\bar{\bar{A}}) = A$$
$$A + \bar{A} = S$$
$$A\bar{A} = \varnothing$$
$$\bar{A} \subset \bar{B}, \qquad \text{if } B \subset A$$
$$\bar{A} = \bar{B}, \qquad \text{if } A = B$$

Two additional relations that are usually referred to as *DeMorgan's laws* are

$$\overline{(A + B)} = \bar{A}\bar{B}$$
$$\overline{(AB)} = \bar{A} + \bar{B}$$

Differences.   The difference of two sets, $A - B$, is a set consisting of the elements of A that are not in B.   This is shown in Figure 1-6.   The difference may also be expressed as

$$A - B = A\bar{B} = A - AB$$

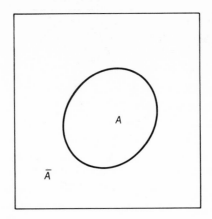

Figure 1-5   The complement of A.

The notation $(A - B)$ is often read as "A take away B."   The following results are also apparent from the Venn diagram:

$$(A - B) + B \neq A$$
$$(A + A) - A = \varnothing$$
$$A + (A - A) = A$$
$$A - \varnothing = A$$
$$A - S = \varnothing$$
$$S - A = \bar{A}$$

Note that when differences are involved, the parentheses *cannot* be omitted.

It is desirable to illustrate all of the above operations with a specific example.   In order to do this, let the elements of the space S be the integers from 1 to 6, as before:

$$S = \{1,2,3,4,5,6\}$$

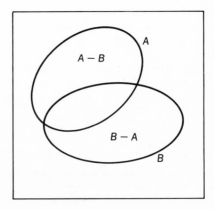

Figure 1-6   The difference of two sets.

and define certain sets as

$$A = \{2,4,6\}, \qquad B = \{1,2,3,4\}, \qquad C = \{1,3,5\}$$

From the definitions just presented, it is clear that

$(A + B) = \{1,2,3,4,6\}, (B + C) = \{1,2,3,4,5\}$
$A + B + C = \{1,2,3,4,5,6\} = S = A + C$
$AB = \{2,4\}, BC = \{1,3\}, AC = \varnothing$
$ABC = \varnothing, \bar{A} = \{1,3,5\} = C, \bar{B} = \{5,6\}$
$\bar{C} = \{2,4,6\} = A, A - B = \{6\}, B - A = \{1,3\}$
$A - C = \{2,4,6\} = A, C - A = \{1,3,5\} = C, B - C = \{2,4\}$
$C - B = \{5\}, (A - B) + B = \{1,2,3,4,6\}$

The student should verify these results.

1-5   The Axiomatic Approach

It is now necessary to relate probability theory to the set concepts that have just been discussed. This relationship is established by defining a *probability space* whose elements are all the outcomes (of a possible set of outcomes) from an experiment. For example, if an experimenter chooses to view the six faces of a die as the possible outcomes, then the probability space associated with throwing a die is the set

$$S = \{1,2,3,4,5,6\}$$

The various subsets of S can be identified with the *events*. For example, in the case of throwing a die, the event $\{2\}$ corresponds to obtaining the outcome 2, while the event $\{1,2,3\}$ corresponds to the outcomes of either 1, or 2, or 3. Since at least one outcome must be obtained on each trial, the space S corresponds to the *certain event* and the empty set $\varnothing$ corresponds to the *impossible event*. Any event consisting of a single element is called an *elementary event*.

The next step is to assign to each event a number called, as before, the *probability of the event*. If the event is denoted as A, the probability of event A is denoted as Pr (A). This number is chosen so as to satisfy the following three conditions or *axioms:*

$$\text{Pr } (A) \geq 0 \tag{1-7}$$
$$\text{Pr } (S) = 1 \tag{1-8}$$
$$\text{If } AB = \varnothing, \text{ then Pr } (A + B) = \text{Pr } (A) + \text{Pr } (B) \tag{1-9}$$

The whole body of probability can be deduced from these axioms. It should be emphasized, however, that axioms are postulates and, as such, it is meaningless to try to prove them. The only possible test of their validity is whether the resulting theory adequately represents the real world. The same is true of any physical theory.

A large number of corollaries can be deduced from these axioms and a few are developed here.   First, since

$$S\varnothing = \varnothing \qquad \text{and} \qquad S + \varnothing = S$$

it follows from (1-9) that

$$\Pr(S + \varnothing) = \Pr(S) = \Pr(S) + \Pr(\varnothing)$$

Hence,

$$\Pr(\varnothing) = 0 \tag{1-10}$$

Next, since

$$A\bar{A} = \varnothing \qquad \text{and} \qquad A + \bar{A} = S$$

it also follows from (1-9) that

$$\Pr(A + \bar{A}) = \Pr(A) + \Pr(\bar{A}) = \Pr(S) = 1 \tag{1-11}$$

From this and from (1-7)

$$\Pr(A) = 1 - \Pr(\bar{A}) \leq 1 \tag{1-12}$$

Therefore, the probability of an event must be a number between 0 and 1.

If A and B are not mutually exclusive, then (1-9) usually does not hold.   A more general result can be obtained, however.   From the Venn diagram of Figure 1-3 it is apparent that

$$A + B = A + \bar{A}B$$

and that A and $\bar{A}B$ are mutually exclusive.   Hence, from (1-9) it follows that

$$\Pr(A + B) = \Pr(A + \bar{A}B) = \Pr(A) + \Pr(\bar{A}B)$$

From the same figure it is also apparent that

$$B = AB + \bar{A}B$$

and that AB and $\bar{A}B$ are mutually exclusive.   From (1-9)

$$\Pr(B) = \Pr(AB + \bar{A}B) = \Pr(AB) + \Pr(\bar{A}B) \tag{1-13}$$

Upon eliminating $\Pr(\bar{A}B)$, it follows that

$$\Pr(A + B) = \Pr(A) + \Pr(B) - \Pr(AB) \leq \Pr(A) + \Pr(B) \tag{1-14}$$

which is the desired result.

Now that the formalism of the axiomatic approach has been established, it is desirable to look at the problem of constructing probability spaces.   First consider the case of throwing a single die and the associated probability space of $S = \{1,2,3,4,5,6\}$.   The elementary events are simply the integers associated with the upper face of the die and these are clearly mutually exclusive.   If the elementary events are *assumed* to be equally probable, then the probability associated with each is simply

$$\text{Pr } \{\alpha_i\} = \frac{1}{6}, \qquad \alpha_i = 1, 2, \ldots, 6$$

Note that this assumption is consistent with the relative-frequency approach, but within the framework of the axiomatic approach it is only an assumption, and any number of other assumptions could have been made.

For this same probability space, consider the event $A = \{1,3\} = \{1\} + \{3\}$. From (1-9)

$$\text{Pr } (A) = \text{Pr } \{1\} + \text{Pr } \{3\} = \frac{1}{6} + \frac{1}{6} = \frac{1}{3}$$

and this can be interpreted as the probability of throwing *either* a 1 or a 3. A somewhat more complex situation arises when $A = \{1,3\}$, $B = \{3,5\}$ and it is desired to determine $\text{Pr } (A + B)$. Since A and B are not mutually exclusive, the result of (1-14) must be used. From the calculation above, it is clear that $\text{Pr } (A) = \text{Pr } (B) = \frac{1}{3}$. However, since $AB = \{3\}$, an elementary event, it must be that $\text{Pr } (AB) = \frac{1}{6}$. Hence, from (1-14)

$$\text{Pr } (A + B) = \text{Pr } (A) + \text{Pr } (B) - \text{Pr } (AB) = \frac{1}{3} + \frac{1}{3} - \frac{1}{6} = \frac{1}{2}$$

An alternative approach is to note that $A + B = \{1,3,5\}$ which is composed of three mutually exclusive elementary events. Using (1-9) twice leads immediately to

$$\text{Pr } (A + B) = \text{Pr } \{1\} + \text{Pr } \{3\} + \text{Pr } \{5\} = \frac{1}{6} + \frac{1}{6} + \frac{1}{6} = \frac{1}{2}$$

Note that this can be interpreted as the probability of *either* A occurring or B occurring *or both* occurring.

Exercise 1-5

For the probability space of a six-sided die with events $A = \{1\}$, $B = \{odd\}$, $C = \{even\}$ find the probabilities of (a) $\text{Pr } (A + B)$, (b) $\text{Pr } (AB)$, (c) $\text{Pr } (A + C)$, (d) $\text{Pr } (BC)$.

Answer: $0, \frac{2}{3}, \frac{1}{2}, \frac{1}{6}$

1-6   Conditional Probability

The concept of conditional probability was introduced in Section 1-3 on the basis of the relative frequency of one event when another event is specified to have occurred. In the axiomatic approach, conditional probability is a defined quantity.   If an event B is assumed to have a nonzero probability, then the conditional probability of an event A, given B, is defined as

$$\Pr\left(A|B\right) = \frac{\Pr\left(AB\right)}{\Pr\left(B\right)} \qquad \Pr\left(B\right) > 0 \qquad \textbf{(1-15)}$$

where $\Pr\left(AB\right)$ is the probability of the event AB.   In the previous discussion, the numerator of (1-15) was written as $\Pr\left(A,B\right)$ and was called the joint probability of events A and B.   This interpretation is still correct if A and B are elementary events, but in the more general case the proper interpretation must be based on the set theory concept of the product, AB, of two sets.   Obviously, if A and B are mutually exclusive, then AB is the empty set and $\Pr\left(AB\right) = 0$.   On the other hand, if A is contained in B (that is, $A \subset B$), then $AB = A$ and

$$\Pr\left(A|B\right) = \frac{\Pr\left(A\right)}{\Pr\left(B\right)} \geq \Pr\left(A\right)$$

Finally, if $B \subset A$, then $AB = B$ and

$$\Pr\left(A|B\right) = \frac{\Pr\left(B\right)}{\Pr\left(B\right)} = 1$$

In general, however, when neither $A \subset B$ nor $B \subset A$, nothing can be asserted regarding the relative magnitudes of $\Pr\left(A\right)$ and $\Pr\left(A|B\right)$.

So far it has not yet been shown that conditional probabilities are really probabilities in the sense that they satisfy the basic axioms.   In the relative-frequency approach they are clearly probabilities in that they could be defined as ratios of the numbers of favorable occurrences to the total number of trials, but in the axiomatic approach conditional probabilities are defined quantities; hence it is necessary to verify independently their validity as probabilities.

The first axiom is

$$\Pr\left(A|B\right) \geq 0$$

and this is obviously true from the definition (1-15) since both numerator and denominator are positive numbers.   The second axiom is

$$\Pr\left(S|B\right) = 1$$

and this is also apparent since $B \subset S$ so that $SB = B$ and $\Pr\left(SB\right) = \Pr\left(B\right)$.   In order to verify that the third axiom holds, consider another event, C, such that $AC = \varnothing$ (that is, A and C are mutually exclusive). Then

$$\Pr\left[(A + C)B\right] = \Pr\left[(AB) + (CB)\right] = \Pr\left(AB\right) + \Pr\left(CB\right)$$

since (AB) and (CB) are also mutually exclusive events and (1-9) holds for such events.   So, from (1-15)

$$\Pr\left[(A+C)|B\right] = \frac{\Pr\left[(A+C)B\right]}{\Pr(B)} = \frac{\Pr(AB)}{\Pr(B)} + \frac{\Pr(CB)}{\Pr(B)} = \Pr(A|B)$$
$$+ \Pr(C|B)$$

Thus the third axiom does hold, and it is now clear that conditional probabilities are valid probabilities in every sense.

Before extending the topic of conditional probabilities, it is desirable to consider an example in which the events are not elementary events. Let the experiment be the throwing of a single die and let the outcomes be the integers from 1 to 6.   Then define event A as $A = \{1,2\}$, that is, the occurrence of a 1 or a 2.   From previous considerations it is clear that $\Pr(A) = \frac{1}{6} + \frac{1}{6} = \frac{1}{3}$.   Define B as the event of obtaining an even number.   That is, $B = \{2,4,6\}$ and $\Pr(B) = \frac{1}{2}$ since it is composed of three elementary events.   The event AB is $AB = \{2\}$, from which $\Pr(AB) = \frac{1}{6}$.   The conditional probability, $\Pr(A|B)$, is now given by

$$\Pr(A|B) = \frac{\Pr(AB)}{\Pr(B)} = \frac{\frac{1}{6}}{\frac{1}{2}} = \frac{1}{3}$$

This indicates that the conditional probability of throwing a 1 or a 2, given that the outcome is even, is $\frac{1}{3}$.

On the other hand, suppose it is desired to find the conditional probability of throwing an even number given that the outcome was a 1 or a 2. This is

$$\Pr(B|A) = \frac{\Pr(AB)}{\Pr(A)} = \frac{\frac{1}{6}}{\frac{1}{3}} = \frac{1}{2}$$

a result that is intuitively correct.

One of the uses of conditional probability is in the evaluation of *total probability*.   Suppose there are n mutually exclusive events A_1, $A_2, \ldots, A_n$ and an arbitrary event B as shown in the Venn diagram of Figure 1-7.   The events A_i occupy the entire space, S, so that

$$A_1 + A_2 + \cdots + A_n = S \tag{1-16}$$

Since A_i and A_j $(i \neq j)$ are mutually exclusive, it follows that BA_i and BA_j are also mutually exclusive.   Further,

$$B = B(A_1 + A_2 + \cdots + A_n) = BA_1 + BA_2 + \cdots + BA_n$$

because of (1-16).   Hence, from (1-9),

$$\Pr(B) = \Pr(BA_1) + \Pr(BA_2) + \cdots + \Pr(BA_n) \tag{1-17}$$

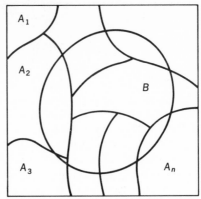

Figure 1-7   Venn diagram for total probability.

But from (1-15)

$$\text{Pr } (BA_i) = \text{Pr } (B|A_i) \text{ Pr } (A_i)$$

Substituting into (1-17) yields

$$\text{Pr } (B) = \text{Pr } (B|A_1) \text{ Pr } (A_1) + \text{Pr } (B|A_2) \text{ Pr } (A_2) + \cdots$$
$$+ \text{ Pr } (B|A_n) \text{ Pr } (A_n) \quad \textbf{(1-18)}$$

The quantity Pr (B) is the *total probability* and is expressed in (1-18) in terms of its various conditional probabilities.

An example serves to illustrate an application of total probability. Consider a resistor carrousel containing six bins. Each bin contains an assortment of resistors as shown in Table 1-3.

Table 1-3   RESISTANCE VALUES

			BIN NUMBERS				
Ohms	1	2	3	4	5	6	Total
10 Ω	500	0	200	800	1200	1000	3700
100 Ω	300	400	600	200	800	0	2300
1000 Ω	200	600	200	600	0	1000	2600
Total	1000	1000	1000	1600	2000	2000	8600

If one of the bins is selected at random,[1] and a single resistor drawn from that bin at random, what is the probability that the resistor chosen will

---

[1] The phrase "at random" is usually interpreted to mean "with equal probability."

be 10 Ω?   The A_i events in (1-18) can be associated with the bin chosen so that

$$Pr(A_i) = \frac{1}{6}, \quad i = 1, 2, 3, 4, 5, 6$$

since it is *assumed* that the choices of bins are equally likely.   The event B is the selection of a 10 Ω resistor and the conditional probabilities can be related to the *numbers* of such resistors in each bin.   Thus

$$Pr(B|A_1) = \frac{500}{1000} = \frac{1}{2} \qquad Pr(B|A_2) = \frac{0}{1000} = 0$$

$$Pr(B|A_3) = \frac{200}{1000} = \frac{2}{10} \qquad Pr(B|A_4) = \frac{800}{1600} = \frac{1}{2}$$

$$Pr(B|A_5) = \frac{1200}{2000} = \frac{6}{10} \qquad Pr(B|A_6) = \frac{1000}{2000} = \frac{1}{2}$$

Hence, from (1-18) the total probability of selecting a 10 Ω resistor is

$$Pr(B) = \frac{1}{2} \times \frac{1}{6} + 0 \times \frac{1}{6} + \frac{2}{10} \times \frac{1}{6} + \frac{1}{2} \times \frac{1}{6} + \frac{6}{10} \times \frac{1}{6} + \frac{1}{2} \times \frac{1}{6}$$
$$= 0.3833$$

It is worth noting that the relative frequency concept is used in assigning values to the conditional probabilities above, but that the basic relationship expressed by (1-18) is derived from the axiomatic approach.

The probabilities $Pr(A_i)$ in (1-18) are often referred to as *a priori probabilities* because they are the ones that describe the probabilities of the events A_i *before* any experiment is performed.   *After* an experiment is performed, and event B observed, the probabilities that describe the events A_i are the *conditional probabilities* $Pr(A_i|B)$.   These probabilities may be expressed in terms of those already discussed by rewriting (1-15) as

$$Pr(A_iB) = Pr(A_i|B) Pr(B) = Pr(B|A_i) Pr(A_i)$$

The last form in the above is obtained by simply interchanging the roles of B and A_i.   The second equality may now be written

$$Pr(A_i|B) = \frac{Pr(B|A_i) Pr(A_i)}{Pr(B)}, \quad Pr(B) \neq 0 \qquad \text{(1-19)}$$

into which (1-18) may be substituted to yield

$$Pr(A_i|B) = \frac{Pr(B|A_i) Pr(A_i)}{Pr(B|A_1) Pr(A_1) + \cdots + Pr(B|A_n) Pr(A_n)} \qquad \text{(1-20)}$$

The conditional probability $Pr(A_i|B)$ is often called the *a posteriori probability* because it applies *after* the experiment is performed; and either (1-19) or (1-20) is referred to as *Bayes' theorem*.

The *a posteriori* probability may be illustrated by continuing the example just discussed. Suppose the resistor that is chosen from the carrousel is found to be a 10-Ω resistor. What is the probability that it came from bin three? Since B is still the event of selecting a 10-Ω resistor, the conditional probabilities $\Pr(B|A_i)$ are the same as tabulated before. Furthermore, the *a priori* probabilities are still $\frac{1}{6}$. Thus, from (1-19), and the previous evaluation of $\Pr(B)$,

$$\Pr(A_3|B) = \frac{\left(\frac{2}{10}\right)\left(\frac{1}{6}\right)}{0.3833} = 0.0869$$

This is the probability that the 10-Ω resistor, chosen at random came from bin three.

Exercise 1-6

Using the data of Table 1-3, find the probability (a) that a 100-Ω resistor that is drawn came from bin two or (b) that a 1000-Ω resistor that is drawn came from bin two.

Answer: 0.321, 0.219

1-7   Independence

The concept of statistical independence is a very important one in probability. It was introduced in connection with the relative-frequency approach by considering two trials of an experiment, such as tossing a coin, in which it is clear that the second trial cannot depend upon the outcome of the first trial in any way. Now that a more general formulation of events is available, this concept can be extended. The basic definition is unchanged, however. It is

Two events, A and B, are independent if and only if
$$\Pr(AB) = \Pr(A)\Pr(B) \tag{1-21}$$

In many physical situations, independence of events is *assumed* because there is no apparent physical mechanism by which one event can depend upon the other. In other cases the assumed probabilities of the elementary events lead to independence of other events defined from these. In such cases, independence may not be obvious, but can be established from (1-21).

The concept of independence can also be extended to more than two events. For example, with three events, the conditions for independence are

$$\Pr(A_1A_2) = \Pr(A_1)\Pr(A_2) \qquad \Pr(A_1A_3) = \Pr(A_1)\Pr(A_3)$$
$$\Pr(A_2A_3) = \Pr(A_2)\Pr(A_3) \qquad \Pr(A_1A_2A_3) = \Pr(A_1)\Pr(A_2)\Pr(A_3)$$

Note that *four* conditions must be satisfied, and that pairwise independence is *not sufficient* for the entire set of events to be mutually independent. In general, if there are n events, it is necessary that

$$\Pr (A_i A_j \cdot \cdot \cdot A_k) = \Pr (A_i) \Pr (A_j) \cdot \cdot \cdot \Pr (A_k) \qquad \text{(1-22)}$$

for every set of integers less than or equal to n. This implies that $2^n - (n + 1)$ equations of the form (1-22) are required to establish the independence of n events.

One important consequence of independence is a special form of (1-14), which stated

$$\Pr (A + B) = \Pr (A) + \Pr (B) - \Pr (AB)$$

If A and B are independent events, this becomes

$$\Pr (A + B) = \Pr (A) + \Pr (B) - \Pr (A) \Pr (B) \qquad \text{(1-23)}$$

Another result of independence is

$$\Pr [A_1(A_2 + A_3)] = \Pr (A_1) \Pr (A_2 + A_3) \qquad \text{(1-24)}$$

if A_1, A_2, and A_3 are all independent. This is *not* true if they are only independent in pairs. In general, if A_1, A_2, . . . , A_n are independent events, then any one of them is independent of any event formed by sums, products, and complements of the others.

Examples of physical situations that illustrate independence are most often associated with two or more trials of an experiment. It is possible, of course, to define sets associated with a single trial that are independent, but these sets may not represent any physical situation. For example, consider throwing a single die and define two events as $A = \{1,2,3\}$ and $B = \{3,4\}$. From previous results it is clear that $\Pr (A) = \frac{1}{2}$ and $\Pr (B) = \frac{1}{3}$. The event (AB) contains a single element $\{3\}$; hence, $\Pr (AB) = \frac{1}{6}$. Thus, it follows that

$$\Pr (AB) = \frac{1}{6} = \Pr (A) \Pr (B) = \frac{1}{2} \cdot \frac{1}{3} = \frac{1}{6}$$

and events A and B are independent, although the physical significance of this is not intuitively clear. The next section considers situations in which there is more than one experiment, or more than one trial of a given experiment, and that discussion will help clarify the matter.

1-8   Combined Experiments

In the discussion of probability presented thus far, the probability space, S, was associated with a single experiment. This concept is too restrictive to deal with many realistic situations, so it is necessary to generalize it somewhat.   Consider a situation in which *two* experiments are performed.   For example, one experiment might be throwing a die and the other one tossing a coin.   It is then desired to find the probability that the outcome is, say, a "3" on the die and a "tail" on the coin.   In other situations the second experiment might be simply a repeated trial of the first experiment.   The two experiments, taken together, form a *combined experiment*, and it is now necessary to find the appropriate probability space for it.

Let one experiment have a space S_1 and the other experiment a space S_2.   Designate the elements of S_1 as

$$S_1 = \{\alpha_1, \alpha_2, \ldots, \alpha_n\}$$

and those of S_2 as

$$S_2 = \{\beta_1, \beta_2, \ldots, \beta_m\}$$

Then form a new space, called the *cartesian product space*, whose elements are all the ordered pairs (α_1, β_1), (α_1, β_2), $\ldots$, (α_i, β_j), $\ldots$, (α_n, β_m). Thus, if S_1 has n elements and S_2 has m elements, the cartesian product space has mn elements.   The cartesian product space may be denoted as

$$S = S_1 \times S_2$$

to distinguish it from the previous product or intersection discussed in Section 1-4.

As an illustration of the cartesian product space for combined experiments, consider the die and the coin discussed above.   For the die the space is

$$S_1 = \{1,2,3,4,5,6\}$$

while for the coin it is

$$S_2 = \{H,T\}$$

Thus, the cartesian product space has 12 elements and is

$$S = S_1 \times S_2 = \{(1,H), (1,T), (2,H), (2,T), (3,H), (3,T), (4,H),$$
$$(4,T), (5,H), (5,T), (6,H), (6,T)\}$$

It is now necessary to define the events of the new probability space. If A_1 is a subset considered to be an event in S_1, and A_2 is a subset considered to be an event in S_2, then $A = A_1 \times A_2$ is an event in S.   For example, in the above illustration let $A_1 = \{1,3,5\}$ and $A_2 = \{H\}$.   The event A corresponding to these is

$$A = A_1 \times A_2 = \{(1,H), (3,H), (5,H)\}$$

In order to specify the probability of event A, it is necessary to consider whether the two experiments are independent; the only cases discussed here are those in which they are independent.   In such cases the probability in the product space is simply the products of the probabilities in the original spaces.   Thus, if $\Pr(A_1)$ is the probability of event

A_1 in space S_1, and Pr (A_2) is the probability of A_2 in space S_2, then the probability of event A in space S is

$$\text{Pr } (A) = \text{Pr } (A_1 \times A_2) = \text{Pr } (A_1) \text{ Pr } (A_2) \qquad \textbf{(1-25)}$$

This result may be illustrated by data from the above example. From previous results, Pr $(A_1) = \frac{1}{6} + \frac{1}{6} + \frac{1}{6} = \frac{1}{2}$ when $A_1 = \{1,3,5\}$ and Pr $(A_2) = \frac{1}{2}$ when $A_2 = \{H\}$. Thus, the probability of getting an *odd number* on the die and a *head* on the coin is

$$\text{Pr } (A) = \left(\frac{1}{2}\right)\left(\frac{1}{2}\right) = \frac{1}{4}$$

It is possible to generalize the above ideas in a straightforward manner to situations in which there are more than two experiments. However, this will be done only for the more specialized situation of repeating the same experiment an arbitrary number of times.

1-9 Bernoulli Trials

The situation considered here is one in which the same experiment is repeated n times and it is desired to find the probability that a particular event occurs exactly k of these times. For example, what is the probability that exactly four heads will be observed when a coin is tossed ten times? Such repeated experiments are referred to as *Bernoulli trials*.

Consider some experiment for which the event A has a probability Pr $(A) = p$. Hence, the probability that the event does not occur is Pr $(\bar{A}) = q$, where $p + q = 1$.[2] Then repeat this experiment n times and assume that the trials are independent; that is, that the outcome of any one trial does not depend in any way upon the outcomes of any previous (or future) trials. Next determine the probability that event A occurs exactly k times in some specific order, say in the first k trials and none thereafter. Because the trials are independent, the probability of this event is

$$\underbrace{\text{Pr } (A) \text{ Pr } (A) \cdots \text{ Pr } (A)}_{k \text{ of these}} \underbrace{\text{Pr } (\bar{A}) \text{ Pr } (\bar{A}) \cdots \text{ Pr } (\bar{A})}_{n - k \text{ of these}} = p^k q^{n-k}$$

However, there are many other ways in which exactly k events could occur because they can arise in any order. Furthermore, because of the independence, all of these other orders should have exactly the same probability as the one specified above. Hence, the event that A occurs

---

[2] The only justification for changing the notation from Pr (A) to p and from Pr $(\bar{A})$ to q is that the p and q notation is traditional in discussing Bernoulli trials and most of the literature uses it.

k times in any order is the sum of the mutually exclusive events that A occurs k times in some specific order, and thus, the probability that A occurs k times is simply the above probability for a particular order multiplied by the number of different orders that can occur.

The number of different orders that can arise is simply the number of ways of taking k objects at a time from a set of n objects.   From the theory of combinations, this number is known to be the binomial coefficient defined by[3]

$$\binom{n}{k} = \frac{n!}{k!(n-k)!}$$ **(1-26)**

For example, if $n = 4$ and $k = 2$, then

$$\binom{n}{k} = \frac{4!}{2!2!} = 6$$

and there are six different sequences in which the event A occurs exactly twice.   These can be enumerated easily as

$$A A \bar{A} \bar{A},\ A \bar{A} A \bar{A},\ A \bar{A} \bar{A} A,\ \bar{A} A \bar{A} A,\ \bar{A} A A \bar{A},\ \bar{A} \bar{A} A A$$

It is now possible to write the desired probability of A occurring k times as

$$p_n(k) = \text{Pr}\ \{A \text{ occurs } k \text{ times}\} = \binom{n}{k} p^k q^{n-k}$$ **(1-27)**

As an illustration of a possible application of this result, consider a digital computer in which the binary digits (0 or 1) are organized into "words" of 32 digits each.   If there is a probability of 10^{-3} that any one binary digit is incorrectly read, what is the probability that there is one error in an entire word?   For this case, $n = 32$, $k = 1$, and $p = 10^{-3}$. Hence,

$$\text{Pr}\ \{\text{one error in a word}\} = p_{32}(1) = \binom{32}{1} (10^{-3})^1 (0.999)^{31}$$

$$= 32(0.999)^{31}(10^{-3}) \simeq 0.031$$

It is also possible to use (1-27) to find the probability that there will be *no* error in a word.   For this, $k = 0$ and $\binom{n}{0} = 1$.   Thus,

$$\text{Pr}\ \{\text{no error in a word}\} = p_{32}(0) = \binom{32}{0} (10^{-3})^0 (0.999)^{32}$$

$$= (0.999)^{32} \simeq 0.9685$$

There are many other practical applications of Bernoulli trials.   For

---

[3] A table of binominal coefficients is given in Appendix C.

example, if a system has n components and there is a probability p that any one of them will fail, the probability that one and only one component will fail is

$$\text{Pr \{one failure\}} = p_n(1) = \binom{n}{1} pq^{(n-1)}$$

In some cases one may be interested in determining the probability that event A occurs at least k times, or the probability that it occurs no more than k times. These probabilities may be obtained by simply adding the probabilities of all the outcomes that are included in the desired event. For example, if a coin is tossed four times, what is the probability of obtaining at least two heads? For this case, $p = q = \frac{1}{2}$ and $n = 4$. From (1-27) the probability of getting two heads (that is, $k = 2$) is

$$p_4(2) = \binom{4}{2} \left(\frac{1}{2}\right)^2 \left(\frac{1}{2}\right)^2 = (6) \left(\frac{1}{4}\right)\left(\frac{1}{4}\right) = \frac{3}{8}$$

Similarly the probability of three heads is

$$p_4(3) = \binom{4}{3} \left(\frac{1}{2}\right)^3 \left(\frac{1}{2}\right)^1 = (4) \left(\frac{1}{8}\right)\left(\frac{1}{2}\right) = \frac{1}{4}$$

and the probability of four heads is

$$p_4(4) = \binom{4}{4} \left(\frac{1}{2}\right)^4 \left(\frac{1}{2}\right)^0 = (1) \left(\frac{1}{16}\right) (1) = \left(\frac{1}{16}\right)$$

Hence, the probability of getting at least two heads is

$$\text{Pr \{at least two heads\}} = p_4(2) + p_4(3) + p_4(4) = \frac{3}{8} + \frac{1}{4} + \frac{1}{16} = \frac{11}{16}$$

The general formulation of problems of this kind can be expressed quite easily, but there are several different situations that arise. These may be tabulated as follows:

$$\text{Pr \{A occurs \textit{less} than k times in n trials\}} = \sum_{i=0}^{k-1} p_n(i)$$

$$\text{Pr \{A occurs \textit{more} than k times in n trials\}} = \sum_{i=k+1}^{n} p_n(i)$$

$$\text{Pr \{A occurs \textit{no more} than k times in n trials\}} = \sum_{i=0}^{k} p_n(i)$$

$$\text{Pr \{A occurs \textit{at least} k times in n trials\}} = \sum_{i=k}^{n} p_n(i)$$

A final comment in regard to Bernoulli trials has to do with evaluating

$p_n(k)$ when n is large. Since the binomial coefficients and the large powers of p and q become difficult to evaluate in such cases, often it is necessary to seek simpler, but approximate, ways of carrying out the calculation. One such approximation, known as the *DeMoivre-Laplace theorem*, is useful if $npq \gg 1$ and if $|k - np|$ is on the order of or less than $\sqrt{npq}$. This approximation is

$$p_n(k) = \binom{n}{k} p^k q^{n-k} \simeq \frac{1}{\sqrt{2\pi npq}}\, \epsilon^{-(k-np)^2/2npq} \qquad \textbf{(1-28)}$$

The DeMoivre-Laplace theorem has additional significance when continuous probability is considered in a subsequent chapter. However, a simple illustration of its utility in discrete probability is worthwhile. Suppose a coin is tossed 100 times and it is desired to find the probability of k heads, where k is in the vicinity of 50. Since $p = q = \frac{1}{2}$ and $n = 100$, (1-28) yields

$$p_n(k) \simeq \frac{1}{\sqrt{50\pi}}\, \epsilon^{-(k-50)^2/50}$$

for k values ranging (roughly) from 40 to 60. This is obviously much easier to evaluate than trying to find the binomial coefficient $\binom{100}{k}$ for the same range of k values.

Exercise 1-9

In a particular state it has been found that during a typical 60 hour weekend about 15 fatal accidents occur. Assuming that the probability of an accident occurring in any one-hour period is the same as any other one-hour period determine the probability of (a) no accidents in a six-hour period, (b) 5 accidents in a six-hour period, (c) no accidents over a weekend.

Answers: $9/2048$, 3.2×10^{-8}, $729/4096$

■ PROBLEMS

1-1 If a single die is rolled, determine the probability of each of the following events:

a. Obtaining the number 3.
b. Obtaining a number less than three.
c. Obtaining an odd number.

1-2   If two dice are rolled, determine the probability of each of the following events:

a. Obtaining a sum of 7.
b. Obtaining a sum greater than 5.
c. Obtaining an odd sum.

1-3   A company manufactures small electric motors having power ratings of $\frac{1}{5}$, $\frac{1}{4}$ or $\frac{1}{3}$ horsepower and designed for operation with 120 V single-phase ac, 240 V single phase ac or 120 V dc.   The motor types can only be distinguished by the differences in their nameplates.   A distributor has on hand 2500 motors in the quantities shown in the table below.

HORSEPOWER	120 V AC	240 V AC	120 V DC	TOTALS
$\frac{1}{5}$	250	750	500	1500
$\frac{1}{4}$	150	250	250	650
$\frac{1}{3}$	100	250	0	350
Totals	500	1250	750	2500

One motor is discovered without a nameplate.   Determine the probability of each of the following events:

a. The motor has a power rating of $\frac{1}{4}$ hp.
b. The motor utilizes an input voltage of 120 V dc.
c. The motor is rated at $\frac{1}{4}$ hp *and* utilizes an input voltage of 120 V dc.
d. The motor is rated at $\frac{1}{3}$ hp *and* utilizes an input voltage of 120 V dc.

1-4   In problem 1-3, assume that 10 percent of the marked 120 V single-phase ac motors are mislabeled and that 2 percent of the motors marked 120 V dc are mislabeled.

a. What is the probability that a motor picked at random is mislabeled?
b. What is the probability that a motor picked at random from those marked $\frac{1}{4}$ hp is mislabeled?
c. What is the probability that a motor picked at random is $\frac{1}{5}$ hp and mislabeled?

1-5   Compute the following probabilities:

a. Cutting a deck of cards five consecutive times and getting an ace on at least three of the cuts.
b. Having at least three aces dealt in a five card poker hand.
c. Drawing one card to a flush.

1-6   Prove that a space S containing n elements has 2^n subsets.

1-7 A space S is defined as

$$S = \{1,2,3,4,5,6,7,8,9,10\}$$

and three subsets of this space as

$$A = \{1,2,3\}, \; B = \{1,3,5,7,9\}, \; C = \{2,4,6,8,10\}.$$

Find:

$A + B$	ABC	$\overline{BC}$
$B + C$	$\bar{A}$	$A - C$
$A + C$	$\bar{B}$	$C - A$
AB	$\bar{C}$	$A - B$
AC	$\bar{A}B$	$(A - B) + B$
BC	$A\bar{B}$	$(A - B) + C$

1-8 Draw and label the Venn diagram for problem 1-4.

1-9 Show that for any 3 events, A, B, C, defined on a probability space, the probability that at least 1 of the events will occur is given by

$$\Pr(A + B + C) = \Pr(A) + \Pr(B) + \Pr(C)$$
$$- \Pr(AB) - \Pr(AC) - \Pr(BC) + \Pr(ABC).$$

1-10 A certain typist sometimes makes mistakes by hitting a key to the right or left of the intended key, each with a probability of 0.05. The letters E, R, and T are adjacent to one another on the keyboard, and in English they occur with the probabilities of $\Pr(E) = 0.1031$, $\Pr(R) = 0.0484$, and $\Pr(T) = 0.0796$. What is the probability of the letter R in the copy typed by this typist from an English original?

1-11 In a certain communication system the message is coded into the binary digits 0 and 1. After coding, the probability of a 0 being transmitted is 0.45, while the probability of a 1 is 0.55. In the communication channel the probability of a transmitted 0 being distorted into a 1 at the receiver is 0.1, while the probability of a 1 being distorted into a 0 is 0.2. Find:

a. The probability that a received 0 was transmitted as a 0.
b. The probability that a received 1 was transmitted as a 1.

1-12 In playing an opponent of equal ability, what is more probable:

a. To win 3 games out of 4, or to win 5 games out of 8?
b. To win at least 3 games out of 4, or to win at least 5 games out of 8?

1-13 A certain football receiver is able to catch $\frac{2}{3}$ of the passes thrown to him. He must catch three passes for his team to win the game. The quarterback throws the ball to the receiver five times.

a. What is the probability that the receiver will drop the ball all five times?
b. What is the probability that the receiver will win the game for his team?

1-14 A multichannel microwave link is to provide telephone communication to a remote community of 10 subscribers, each of whom uses the link 25 percent of the time during peak hours. How many channels are needed to make the link available during peak hours to:

a. Ninety percent of the subscribers all the time.
b. All of the subscribers 90 percent of the time.

■ REFERENCES

All of the following texts provide coverage of the topics discussed in Chapter 1. Particularly useful and readily understandable discussions are contained in Beckmann, Drake, Gnedenko and Khinchin, Lanning and Battin, and Parzen.

Beckmann, P., *Elements of Applied Probability Theory*. New York: Harcourt, Brace and World, Inc., 1968.

This book provides coverage of much of the material discussed in the first six chapters of the present text. The mathematical level is essentially the same as the present text but the point of view is often different, thereby providing useful amplifications or extensions of the concepts being considered. A number of interesting examples are worked out in the text.

Davenport, W. B., Jr., and W. L. Root, *Introduction to Random Signals and Noise*. New York: McGraw-Hill, Inc., 1958.

This is a graduate level text dealing with the application of probabilistic methods to the analysis of communication systems. The treatment is at an appreciably more advanced mathematical level than the present text and will require some diligent effort on the part of an undergraduate wishing to read it. However, the effort will be amply rewarded as this is the classic book in its field and is the most frequently quoted reference in this area.

Drake, A. W., *Fundamentals of Applied Probability Theory*. New York: McGraw-Hill, Inc., 1967.

This undergraduate text covers the elementary aspects of probability theory in a clear and readable fashion. The material relates directly to Chapters 1, 2 and 3 of the present text. Of particular interest is the use of exponential (Fourier) transforms of the probability density functions of continuous random variables and Z-transforms of the probability density functions of discrete random variables in place of the classical characteristic function procedure.

Gnedenko, B. Y. and A. Ya. Khinchin, *An Elementary Introduction to the Theory of Probability*.   New York: Dover Publications, Inc., 1962.

This small paperback book was written by two outstanding Russian mathematicians for use in high schools.   It provides a very clear and easily understood introduction to many of the basic concepts of probability that are discussed in Chapters 1 and 2 of the present text.   The mathematical level is quite low and, in fact, does not go beyond simple algebra.   Nevertheless, the subject matter is of fundamental importance and much useful insight into probability theory can be obtained from a study of this book.

Lanning, J. H., Jr. and R. H. Battin, *Random Processes in Automatic Control*.   New York: McGraw-Hill, Inc., 1956.

This book is a graduate level text in the field of automatic control.   However, the first half of the book provides a particularly clear and understandable treatment of probability and random processes at a level that is readily understandable by juniors or seniors in electrical engineering.   A number of topics such as random processes are treated in greater detail than in the present text.   This reference contains material relating to virtually all of the topics covered in the present text although some of the applications considered in later chapters involve more advanced mathematical concepts.

Papoulis, A., *Probability, Random Variables, and Stochastic Processes*.   New York: McGraw-Hill, Inc., 1965.

This is a widely used graduate level text aimed at electrical engineering applications of probability theory.   Virtually all of the topics covered in the present text plus a great many more are included.   The treatment is appreciably more abstract and mathematical than the present text but a wide range of useful examples and results are given.   This book provides the most readily available source for many of these results.

Parzen, E., *Modern Probability Theory and its Applications*.   New York: John Wiley and Sons, Inc., 1960.

This is a standard undergraduate text on the mathematical theory of probability.   The material is clearly presented and many interesting applications of probability are considered in the examples and problems.

Thomas, J., *An Introduction to Statistical Communication Theory*.   New York: John Wiley and Sons, Inc., 1969.

In this graduate level text a broad coverage of probability, random processes, and their applications to the analysis of a variety of system problems is given.   Most of the topics discussed in the present text are covered although sometimes from a more advanced mathematical level.   In addition many other topics are considered.   For students desiring a more detailed look at the techniques of probabilistic analysis this text is well worth studying.   The references given at the ends of the chapters are quite extensive and include journal articles as well as books.

2

Random Variables

2-1 Concept of a Random Variable

The previous chapter deals exclusively with situations in which the number of possible outcomes associated with any experiment is finite. Although it is never stated that the outcomes had to be finite in number (because, in fact, they do *not*), such an assumption is implied and is certainly true for such illustrative experiments as tossing coins, throwing dice, and selecting resistors from bins. There are many other experiments, however, in which the number of possible outcomes is not finite, and it is the purpose of this chapter to introduce ways of describing such experiments in accordance with the concepts of probability already established.

A good way to introduce this type of situation is to consider again the experiment of selecting a resistor from a bin. When mention is made, in the previous chapter, of selecting a 1-Ω resistor, or a 10-Ω resistor, or any other value, the implied meaning is that the selected resistor is labeled 1 Ω or 10 Ω. The actual value of resistance is expected to be close to the labeled value, but might differ from it by some unknown (but measurable) amount. The deviations from the labeled value are due to manufacturing variations and can assume any value within some specified range. Since the actual value of resistance is unknown in advance, it is a *random variable.*

To carry this illustration further, consider a bin of resistors that are all marked 100 Ω. Because of manufacturing tolerances, each of the resistors in the bin will have a slightly different resistance value. Furthermore, there are an infinite number of possible resistance values, so that the experiment of selecting one resistor has an infinite number of possible outcomes. Even if it is known that all of the resistance values lie between 99.99 Ω and 100.01 Ω, there are an infinite number of such values in this range. Thus, if one defines a particular event as the selection of a resistor with a resistance of exactly 100.00 Ω, the probability of this event is actually zero. On the other hand, if one were to define an event as the selection of a resistor having a resistance between 99.9999 Ω and 100.0001 Ω, the probability of this event is nonzero. The actual value of resistance, however, is a random variable that can assume any value in a specified range of values.

It is also possible to associate random variables with time functions, and, in fact, most of the applications that are considered in this text are of this type. Although Chapter 3 will deal exclusively with such random variables and random time functions, it is worth digressing momentarily at this point to note the relationship between the two as it provides an important physical motivation for the present study.

A typical random time function, shown in Figure 2-1 is designated

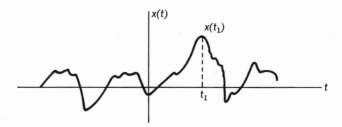

Figure 2-1   A random time function.

as $x(t)$. In a given physical situation, this particular time function is only one of an infinite number of time functions that might have occurred. The collection of all possible time functions that might have been observed belongs to a random process, which will be designated as $\{x(t)\}$. When the probability functions are also specified, this collection is referred to as an ensemble. Any particular member of the ensemble, say $x(t)$, is a sample function, and the value of the sample function at some particular time, say t_1, is a random variable which we call $X(t_1)$ or simply X_1. Thus, $X_1 = x(t_1)$ when $x(t)$ is the particular sample function observed.

A random variable associated with a random process is a considerably more involved concept than the random variable associated with the resistor above. In the first place, there is a different random variable for each instant of time, although there usually is some relation between two random variables corresponding to two different time instants. In the second place, the randomness we are concerned with is the randomness that exists from sample function to sample function throughout the complete ensemble. There may also be randomness from time instant to time instant, but this is not an essential ingredient of a random process. Therefore, the probability description of the random variables being considered here is also the probability description of the random process. However, our initial discussion will concentrate on the random variables and will be extended later to the random process.

From an engineering viewpoint, a random variable is simply a numerical description of the outcome of a random experiment. Recall that the sample space $S = \{\alpha\}$ is the set of all possible outcomes of the experiment. When the outcome is α, the random variable X has a value that we might denote as $X(\alpha)$. From this viewpoint, a random variable is simply a real-valued function defined over the sample space—and in fact the fundamental definition of a random variable is simply as such a function (with a few restrictions needed for mathematical consistency.) For engineering applications, however, it is usually not necessary to consider explicitly the underlying sample space. It is generally only necessary to be able to assign probabilities to various *events* associated with the random variables of interest and these probabilities can often be inferred directly from the physical situation. What events are required for a complete description of the random variable, and how the appropriate probabilities can be inferred, form the subject matter for the rest of this chapter.

If a random variable can assume any value within a specified range (possibly infinite), then it will be designated as a continuous random variable. In the following discussion all random variables will be assumed to be continuous unless stated otherwise. It will be shown, however, that discrete random variables (that is, those assuming one of a countable set of values) can also be treated by exactly the same methods.

2-2 Distribution Functions

In order to consider continuous random variables within the framework of probability concepts discussed in the last chapter, it is necessary to define the *events* to be associated with the probability space. There are many ways in which events might be defined, but the method to be described below is almost universally accepted.

Let X be a random variable as defined above and x be any allowed value of this random variable. The probability distribution function is defined to be the probability of the event that the observed random variable X is less than or equal to the allowed value x. That is,

$$P_X(x) = \Pr (X \leq x)$$

It should be pointed out that using the symbol $P_X(x)$ for the probability distribution function is fairly common in engineering literature but that most mathematics texts use the notation $F_X(x)$ for this function.[1]

Since the probability distribution function is a probability, it must satisfy the basic axioms and must have the same properties as the probabilities discussed in Chapter 1. However, it is also a function of x, the possible values of the random variable X, and as such must generally be defined for all values of x. Thus, the requirement that it be a probability imposes certain constraints upon the functional nature of $P_X(x)$. These may be summarized as follows:

1. $0 \leq P_X(x) \leq 1 \qquad -\infty < x < \infty$
2. $P_X(-\infty) = 0 \qquad P_X(\infty) = 1$
3. $P_X(x)$ is nondecreasing as x increases
4. $\Pr (x_1 < X \leq x_2) = P_X(x_2) - P_X(x_1)$

Some possible distribution functions are shown in Figure 2-2. The sketch in (a) indicates a continuous random variable having possible values ranging from $-\infty$ to ∞ while (b) shows a continuous random variable for which the possible values lie between a and b. The sketch in (c) shows the probability distribution function for a discrete random variable

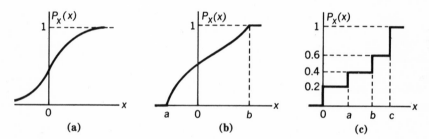

Figure 2-2 Some possible probability distribution functions.

that can assume only four possible values (that is, 0, a, b, or c). In distribution functions of this type it is important to remember that the definition for $P_X(x)$ includes the condition $X = x$ as well as $X < x$. Thus, in Figure 2-2(c), it follows (for example) that $P_X(a) = 0.4$ and not 0.2.

The probability distribution function can also be used to express the probability of the event that the observed random variable X is greater than (but not equal to) x. Since this event is simply the complement

---

[1] The subscript X denotes the random variable while the argument x could equally well be any other symbol. In much of the subsequent discussion it is convenient to suppress the subscript X when no confusion will result. Thus $P_X(x)$ will often be written $P(x)$.

of the event having probability $P_X(x)$ it follows that

$$\text{Pr}\,(X > x) = 1 - P_X(x)$$

As a specific illustration, consider the probability distribution function shown in Figure 2-3. Note that this function satisfies all of the require-

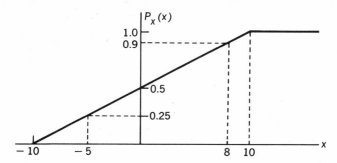

Figure 2-3   A specific probability distribution function.

ments listed above. It is easy to see from the figure that the following statements (among many other possible statements) are true:

$$\text{Pr}\,(X \leq -5) = 0.25$$
$$\text{Pr}\,(X > -5) = 1 - 0.25 = 0.75$$
$$\text{Pr}\,(X > 8) = 1 - 0.9 = 0.1$$
$$\text{Pr}\,(-5 < X \leq 8) = 0.9 - 0.25 = 0.65$$
$$\text{Pr}\,(X \leq 0) = 1 - \text{Pr}\,(X > 0) = 1 - 0.5 = 0.5$$

Exercise 2-2

A particular random variable has a probability distribution function given by

$$P_X(x) = 0 \qquad\qquad -\infty < x \leq -1$$
$$= \frac{1}{2} + \frac{1}{2}x \qquad -1 < x < 1$$
$$= 1 \qquad\qquad 1 \leq x < \infty$$

What is (a) the probability that $X = \frac{1}{4}$, (b) the probability that $X > \frac{3}{4}$, (c) the maximum value that X can assume?

Answer: $\frac{1}{8}, 0, 1$

2-3   Density Functions

Although the distribution function is a complete description of the probability model for a single random variable, it is not the most convenient form for many calculations of interest. For these, it may be preferable to use the derivative of $P(x)$ rather than $P(x)$ itself. This derivative is called the probability density function and, when it exists, it is defined by[2]

$$p_X(x) = \lim_{\epsilon \to 0} \frac{P_X(x + \epsilon) - P_X(x)}{\epsilon} = \frac{dP_X(x)}{dx}$$

The physical significance of the probability density function is best described in terms of the probability element, $p_X(x)\, dx$. This may be interpreted as

$$p_X(x)\, dx = \Pr\,(x < X \leq x + dx) \tag{2-1}$$

Equation (2-1) simply states that the probability element, $p_X(x)\, dx$, is the probability of the event that the random variable X lies in the range of possible values between x and $x + dx$.

Since $p_X(x)$ is a density function and not a probability, it is not necessary that its value be less than 1; it may have any nonnegative value.[3] Its general properties may be summarized as follows:

1. $p_X(x) \geq 0 \qquad -\infty < x < \infty$
2. $\int_{-\infty}^{\infty} p_X(x)\, dx = 1$
3. $P_X(x) = \int_{-\infty}^{x} p_X(u)\, du$
4. $\int_{x_1}^{x_2} p_X(x)\, dx = \Pr\,(x_1 < X \leq x_2)$

As examples of probability density functions, those corresponding to the distribution functions of Figure 2-2 are shown in Figure 2-4.   Note

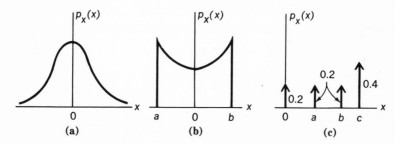

Figure 2-4   Probability density functions corresponding to the distribution functions of Figure 2-2.

particularly that the density function for a discrete random variable consists of a set of delta functions, each having an area equal to the magnitude of the corresponding discontinuity in the distribution function.

---

[2] Again the subscript denotes the random variable and when no confusion results it may be omitted.   Thus $p_X(x)$ will often be written as $p(x)$.

[3] Because $P_X(x)$ is nondecreasing as x increases.

It is also possible to have density functions that contain both a continuous part and one or more delta functions.

There are many different mathematical forms that might be probability density functions, but only a very few of these arise to any significant extent in the analysis of engineering systems. Some of these are considered in subsequent sections and a table containing numerous density functions is given in Appendix B.

A situation that frequently occurs in the analysis of engineering systems is that in which one random variable is functionally related to another random variable whose probability density function is known and it is desired to determine the probability density function of the first random variable. For example, it may be desired to find the probability density function of a power variable when the probability density function of the corresponding voltage or current variable is known. Or it may be desired to find the probability density function after some non-linear operation is performed on a voltage or current. Although a complete discussion of this problem is not necessary here, a few elementary concepts can be presented and will be useful in subsequent discussions.

In order to formulate the mathematical framework, let the random variable Y be a single-valued, real function of another random variable X. Thus, $Y = f(X)$[4], in which it is assumed that the probability density function of X is known and is denoted by $p_X(x)$, and it is desired to find the probability density function of Y, which is denoted by $p_Y(y)$. If it is assumed for the moment that $f(X)$ is a monotonically increasing function of X, then the situation shown in Figure 2-5(a) applies. It is

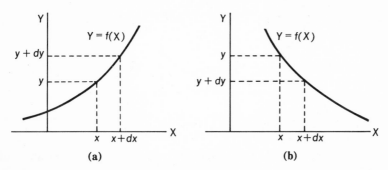

Figure 2-5 Transformation of variables.

clear that whenever the random variable X lies between x and $x + dx$, the random variable Y will lie between y and $y + dy$. Since the probabilities of these events are $p_X(x)\, dx$ and $p_Y(y)\, dy$, one can immediately write

$$p_Y(y)\, dy = p_X(x)\, dx$$

---

[4] This also implies that the possible values of X and Y are related by $y = f(x)$.

from which the desired probability density function becomes

$$p_Y(y) = p_X(x)\frac{dx}{dy} \qquad (2\text{-}2)$$

Or course, in the right side of (2-2), x must be replaced by its corresponding function of y.

When $f(X)$ is a monotonically decreasing function of X, as shown in Figure 2-5(b), a similar result is obtained except that the derivative is negative. Since probability density functions must be positive, and also from the geometry of the figure, it is clear that what is needed in (2-2) is simply the absolute value of the derivative. Hence, for either situation

$$p_Y(y) = p_X(x)\left|\frac{dx}{dy}\right| \qquad (2\text{-}3)$$

There may also be situations in which, for a given Y, $f(X)$ has regions in which the derivative is positive and other regions in which it is negative. In such cases the regions may be considered separately and the corresponding probability densities added. An example of this sort will serve to illustrate the transformation of all random variables.

Let the functional relationship be

$$Y = X^2$$

This is shown in Figure 2-6 and represents, for example, the transformation (except for a scale factor) of a voltage random variable into a power

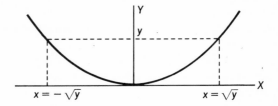

Figure 2-6   The square law transformation.

random variable. Since the derivative, dx/dy, has an absolute value given by

$$\left|\frac{dx}{dy}\right| = \frac{1}{2\sqrt{y}}$$

and since there are two x-values for every y-value ($x = \pm\sqrt{y}$), the desired probability density function is simply

$$p_Y(y) = \frac{1}{2\sqrt{y}}[p_X(\sqrt{y}) + p_X(-\sqrt{y})] \qquad y \geq 0 \qquad (2\text{-}4)$$

Furthermore, since y can never be negative,

$$p_Y(y) = 0 \qquad y < 0$$

Some other applications of random variable transformations will be considered later.

Exercise 2-3

The probability density function of a random variable X has the form $p_X(x) = K[u(x + 1) - u(x - 1)]$, where $u(\cdot)$ is the unit step function. Find the (a) value of K, (b) probability that $X > \frac{3}{4}$, and (c) minimum value of X.

Answer: $\frac{1}{8}, \frac{1}{2}, -1$

2-4   Mean Values and Moments

One of the most important and most fundamental concepts associated with statistical methods is that of finding average values of random variables or functions of random variables. The concept of finding average values for time functions by integrating over some time interval, and then dividing by the length of the interval, is a familiar one to electrical engineers, since operations of this sort are used to find the dc component, the root-mean-square value, or the average power of the time function. Such time averages may also be important for random functions of time, but, of course, have no meaning when considering a single random variable, which is defined as the value of the time function at a single instant of time. Instead, it is necessary to find the average value by integrating over the range of possible values that the random variable may assume. Such an operation is referred to as "ensemble averaging," and the result is the mean value.

Several different notations are in standard use for the mean value but the most common ones in engineering literature are[5]

$$\bar{X} = E[X] = \int_{-\infty}^{\infty} xp(x) \, dx \qquad \text{(2-5)}$$

The symbol $E[X]$ is usually read "the expected value of X" or "the mathematical expectation of X." It will be shown later that in many cases of practical interest the mean value of a random variable is equal to the time average of any sample function from the random process to which the random variable belongs. In such cases, finding the mean value

---

[5] Note that the subscript X has been omitted from $p(x)$ since there is no doubt as to what the random variable is.

of a random voltage or current is equivalent to finding its dc component; this interpretation will be employed here for illustration.

The expected value of any function of x can also be obtained by a similar calculation. Thus,

$$E[f(X)] = \int_{-\infty}^{\infty} f(x)p(x)\,dx \tag{2-6}$$

A function of particular importance is $f(x) = x^n$, since this leads to the general moments of the random variable. Thus,

$$\overline{X^n} = E[X^n] = \int_{-\infty}^{\infty} x^n p(x)\,dx \tag{2-7}$$

By far the most important moments of X are those given by $n = 1$, which is the mean value discussed above, and by $n = 2$, which leads to the mean-square value.

$$\overline{X^2} = E[x^2] = \int_{-\infty}^{\infty} x^2 p(x)\,dx \tag{2-8}$$

The importance of the mean-square value lies in the fact that it may often be interpreted as being equal to the time average of the square of a random voltage or current. In such cases, the mean-square value is proportional to the average power (in a resistor) and its square root is equal to the rms or effective value of the random voltage or current.

It is also possible to define central moments, which are simply the moments of the difference between a random variable and its mean value. Thus, the nth central moment is

$$\overline{(X - \bar{X})^n} = E[(X - \bar{X})^n] = \int_{-\infty}^{\infty} (x - \bar{X})^n p(x)\,dx \tag{2-9}$$

The central moment for $n = 1$ is, of course, zero, while the central moment for $n = 2$ is so important that it carries a special name, the *variance*, and is usually symbolized by σ^2. Thus,

$$\sigma^2 = \overline{(X - \bar{X})^2} = \int_{-\infty}^{\infty} (x - \bar{X})^2 p(x)\,dx \tag{2-10}$$

The variance can also be expressed in an alternative form by using the rules for the expectations of sums; that is,

$$E[X_1 + X_2 + \cdots + X_m] = E[X_1] + E[X_2] + \cdots + E[X_m]$$

Thus,

$$\begin{aligned}
\sigma^2 &= E[(X - \bar{X})^2] = E[X^2 - 2X\bar{X} + \bar{X}^2] \\
&= E[X^2] - 2E[X]\bar{X} + \bar{X}^2 \\
&= \overline{X^2} - 2\bar{X}\bar{X} + \bar{X}^2 = \overline{X^2} - \bar{X}^2
\end{aligned} \tag{2-11}$$

and it is seen that the variance is the difference between the mean-square value and the square of the mean value. The square root of the variance, σ, is known as the *standard deviation*.

In electrical circuits the variance can often be related to the average power (in a resistance) of the ac components of a voltage or current. The square root of the variance would be the value indicated by an ac voltmeter or ammeter of the rms type that does not respond to direct current (because of capacitive coupling, for example).

In order to illustrate some of the above ideas concerning mean values and moments, consider a random variable having a uniform probability density function as shown in Figure 2-7.   A voltage waveform that would

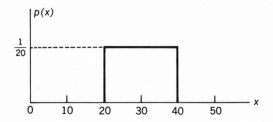

Figure 2-7   A uniform probability density function.

lead to such a probability density function might be a sawtooth waveform that varied linearly between 20 and 40 V.   The appropriate mathematical representation for this density function is

$$p(x) = 0 \qquad -\infty < x \leq 20$$
$$= \frac{1}{20} \qquad 20 < x \leq 40$$
$$= 0 \qquad 40 < x < \infty$$

The mean value of this random variable is obtained by using (2-5). Thus,

$$\bar{X} = \int_{20}^{40} x \left(\frac{1}{20}\right) dx = \frac{1}{20} \cdot \frac{x^2}{2} \Big|_{20}^{40} = \frac{1}{40} (1600 - 400) = 30$$

This value is intuitively the average value of the sawtooth waveform just described.   The mean-square value is obtained from (2-8) as:

$$\overline{X^2} = \int_{20}^{40} x^2 \left(\frac{1}{20}\right) dx = \frac{1}{20} \frac{x^3}{3} \Big|_{20}^{40} = \frac{1}{60} (64 - 8)10^3 = 933.3$$

The variance of the random variable can be obtained from either (2-10) or (2-11).   From the latter,

$$\sigma^2 = \overline{X^2} - (\bar{X})^2 = 933.3 - (30)^{2'} = 33.3$$

On the basis of the assumptions that will be made concerning random processes, if the sawtooth voltage were measured with a dc voltmeter, the reading would be 30 V.   If it were measured with an rms-reading ac voltmeter (which did not respond to dc), the reading would be $\sqrt{33.3}$ V.

Exercise 2-4

A random variable X has a probability density function $p_X(x) = \frac{1}{2}[u(x) - u(x - 2)]$. For the random variable $Y = X^2$, find the (a) mean value, (b) mean-square value, and (c) variance.

Answer: 3.20, 1.33, 1.42

2-5   The Gaussian Random Variable

Of the various density functions that we shall study, the most important by far is the *Gaussian* or *normal* density function.   There are many reasons for its importance, some of which are

1.   It provides a good mathematical model for a great many different physically observed random phenomena.   Furthermore, the fact that it should be a good model can be justified theoretically in many cases.

2.   It is one of the few density functions that can be extended to handle an arbitrarily large number of random variables conveniently.

3.   Linear combinations of Gaussian random variables lead to new random variables that are also Gaussian.   This is not true for most other density functions.

4.   The random process from which Gaussian random variables are derived can be completely specified, in a statistical sense, from a knowledge of all first and second moments only.   This is not true for other processes.

5.   In system analysis, the Gaussian process is often the only one for which a complete statistical analysis can be carried through in either the linear or the nonlinear situation.

The mathematical representation of the Gaussian density function is

$$p(x) = \frac{1}{\sqrt{2\pi}\,\sigma} \exp\left[\frac{-(x - \bar{X})^2}{2\sigma^2}\right] \qquad -\infty < x < \infty \qquad \textbf{(2-12)}$$

where $\bar{X}$ and σ^2 are the mean and variance, respectively.   The corresponding distribution function cannot be written in closed form.   The shapes of the density function and distribution function are shown in Figure 2-8. There are a number of points in connection with these curves that are worth noting.   These are:

1.   There is only one maximum and it occurs at the mean value.

2.   The density function is symmetrical about the mean value.

3.   The width of the density function is directly proportional to the *standard deviation, σ*.   The width of 2σ occurs at the points where the height is 0.607 of the maximum value.   These are also the points of maximum absolute slope.

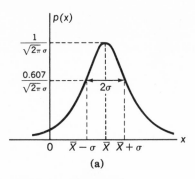 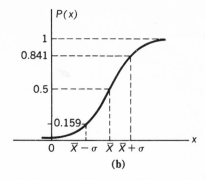

Figure 2-8   The Gaussian random variable: (a) density function and (b) distribution function.

4.   The maximum value of the density function is inversely proportional to the standard deviation σ.   Since the density function has an area of unity, it can be used as a representation of the impulse or delta function by letting σ approach zero.   That is

$$\delta(x - \bar{X}) = \lim_{\sigma \to 0} \frac{1}{\sqrt{2\pi}\,\sigma} \exp\left[\frac{-(x - \bar{X})^2}{2\sigma^2}\right] \qquad \text{(2-13)}$$

This representation of the delta function has an advantage over some others of being infinitely differentiable.

The Gaussian distribution function cannot be expressed in closed form in terms of elementary functions.   It can, however, be expressed in terms of functions that are commonly tabulated.   From the relation between density and distribution functions it follows that the general Gaussian distribution function is

$$P(x) = \int_{-\infty}^{x} p(u)\,du = \frac{1}{\sqrt{2\pi}\,\sigma} \int_{-\infty}^{x} \exp\left[-\frac{(u - \bar{X})^2}{2\sigma^2}\right] du \qquad \text{(2-14)}$$

The function that is usually tabulated is the distribution function for a Gaussian random variable that has a mean value of zero and a variance of unity (that is, $\bar{X} = 0$, $\sigma = 1$).   This distribution function is often designated by $\Phi(x)$ and is defined by

$$\Phi(x) = \frac{1}{\sqrt{2\pi}} \int_{-\infty}^{x} \exp\left[-\frac{u^2}{2}\right] du \qquad \text{(2-15)}$$

By means of a simple change of variable it is easy to show that the general Gaussian distribution function of (2-14) can be expressed in terms of $\Phi(x)$ by

$$P(x) = \Phi\left[\frac{x - \bar{X}}{\sigma}\right] \qquad \text{(2-16)}$$

An abbreviated table of values for $\Phi(x)$ is given in Appendix D.   Since only positive values of x are tabulated, it is frequently necessary to use the additional relationship

$$\Phi(-x) = 1 - \Phi(x) \tag{2-17}$$

Although many of the most useful properties of Gaussian random variables will become apparent only when two or more variables are considered, one that can be mentioned now is the ease with which high-order central moments can be determined. The nth central moment, which was defined in (2-9), can be expressed for a Gaussian random variable as

$$\overline{(X - \bar{X})^n} = 0 \qquad\qquad n \text{ odd}$$
$$= 1 \cdot 3 \cdot 5 \cdots (n - 1)\sigma^n \qquad n \text{ even} \tag{2-18}$$

As an example of the use of (2-18), if $n = 4$, the fourth central moment is $\overline{(X - \bar{X})^4} = 3\sigma^4$. A word of caution should be noted, however. The relation between the nth general moment, $\overline{X^n}$, and the nth central moment is not always as simple as it is for $n = 2$. In the $n = 4$ Gaussian case, for example,

$$\overline{X^4} = 3\sigma^4 + 6\sigma^2(\bar{X})^2 + (\bar{X})^4$$

Before leaving the subject of Gaussian density functions, it is interesting to compare the defining equation, (2-12), with the probability associated with Bernoulli trials for the case of large n as approximated in (1-28). It will be noted that, except for the fact that k and n are integers, the DeMoivre-Laplace approximation has the same form as a Gaussian density function with a mean value of np and a variance of npq. Since the Bernoulli probabilities are discrete, the exact density function for this case is a set of delta functions that increase in number as n increases, and as n becomes large the area of these delta functions follows a Gaussian law.

Another very important result closely related to this is the *central limit theorem*. This famous theorem concerns the *sum* of a large number of independent[6] random variables having the same probability density function. In particular, if the independent random variables are X_1, X_2, . . . , X_n, and the sum is defined by

$$Y = \frac{1}{\sqrt{n}}[X_1 + X_2 + \cdots + X_n]$$

then for large n the density function for Y will approach the Gaussian density function regardless of the density function for the X's. The theorem is also true for more general conditions, but this is not the important aspect here. What is important is to recognize that a great

---

[6] The concept of independence for random variables is defined more precisely in Section 3-3.

many random phenomena that arise in physical situations result from the combined actions of many individual events. This is true for such things as: thermal agitation of electrons in a conductor, shot noise from electrons or holes in a vacuum tube or transistor, atmospheric noise, turbulence in a medium, ocean waves, and many other physical sources of random disturbances. Hence, regardless of the probability density functions of the individual components (and these density functions are usually not even known), one would expect to find that the observed disturbance has a Gaussian density function. The central limit theorem provides a theoretical justification for assuming this, and, in almost all cases experimental measurements bear out the soundness of this assumption.

Exercise 2-5

Find the probability that a Gaussian random variable (a) is less than its mean value and (b) deviates (either direction) from its mean value less than $\pm\sigma$. (c) Find the deviation from the mean (in units of σ) that has a probability of 0.01 of not being exceeded.

Answer: 0.6832, 2.575, 0.5000

2-6   Density Functions Related to Gaussian

The previous section has indicated some of the reasons for the tremendous importance of the Gaussian density function. Still another reason is that there are many other probability density functions, which arise in practical applications, that are related to the Gaussian density function and can be derived from it. The purpose of this section is to list some of these other density functions and indicate the situations under which they arise. They will not all be derived here, since in most cases insufficient background is available, but several of the more important ones will be derived as illustrations of particular techniques.

Distribution of power. When the voltage or current in a circuit is the random variable, the power dissipated in a resistor is also a random variable which is proportional to the square of the voltage or current. The transformation that applies in this case is discussed in Section 2-3 and will be used here to determine the probability density function associated with the power of a Gaussian voltage or current. In particular, let I be the random variable $I(t_1)$ and assume that $p_I(i)$ is Gaussian. The power random variable, W, is then given by

$$W = RI^2$$

and it is desired to find its probability density function $p_W(w)$. By analogy to the result in (2-4), this probability density function may be written as

$$p_W(w) = \frac{1}{2\sqrt{Rw}}\left[p_I\left(\sqrt{\frac{w}{R}}\right) + p_I\left(-\sqrt{\frac{w}{R}}\right)\right] \qquad w \geq 0$$
$$= 0 \qquad\qquad\qquad\qquad w < 0$$

(2-19)

If I is Gaussian and assumed to have zero mean, then

$$p_I(i) = \frac{1}{\sqrt{2\pi}\,\sigma_I}\exp\left[-\frac{i^2}{2\sigma_I{}^2}\right]$$

where $\sigma_I{}^2$ is the variance of I. Hence, σ_I has the physical significance of being the rms value of the current. Furthermore, since the density function is symmetrical, $p_I(i) = p_I(-i)$. Thus, the two terms of (2-19) are identical and the probability density function of the power becomes

$$p_W(w) = \frac{1}{\sigma_I\sqrt{2\pi Rw}}\exp\left[-\frac{w}{2R\sigma_I{}^2}\right] \qquad w \geq 0$$
$$= 0 \qquad\qquad\qquad\qquad w < 0$$

(2-20)

This density function is sketched in Figure 2-9. Straightforward calcu-

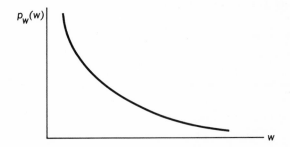

Figure 2-9   Density function for the power of a Gaussian current.

lation indicates that the mean value of the power is

$$\bar{W} = E[RI^2] = R\sigma_I{}^2$$

and the variance of the power is

$$\sigma_W{}^2 = \overline{W^2} - (\bar{W})^2 = E[R^2I^4] - (\bar{W})^2$$
$$= 3R^2\sigma_I{}^4 - (R\sigma_I{}^2)^2 = 2R^2\sigma_I{}^4$$

It may be noted that the probability density function for the power is infinite at $w = 0$; that is, the most probable value of power is zero. This is a consequence of the fact that the most probable value of current is also zero and that the derivative of the transformation (dW/dI) is zero here. It is important to note, however, that there is *not* a delta function in the probability density function.

Rayleigh distribution.   The Rayleigh probability density function arises in several different physical situations.   For example, it will be shown later that the peak values (that is, the *envelope*) of a random voltage or current having a Gaussian probability density function will follow the Rayleigh density function.   The original derivation of this density function (by Lord Rayleigh in 1880) was applied to the envelope of the sum of many sine waves of different frequencies.   It also arises in connection with the errors associated with the aiming of firearms, missiles, and other projectiles, if the errors in each of the two rectangular coordinates have independent Gaussian probability densities.   Thus, if the origin of a rectangular coordinate system is taken to be the target and the error along one axis is X and the error along the other axis is Y, the total miss distance is simply

$$R = \sqrt{X^2 + Y^2}$$

When X and Y are independent Gaussian random variables with zero mean and equal variances, σ^2, the probability density function for R is

$$\begin{aligned} p_R(r) &= \frac{r}{\sigma^2} \exp\left[-\frac{r^2}{2\sigma^2}\right] & r \geq 0 \\ &= 0 & r < 0 \end{aligned} \qquad \text{(2-21)}$$

This is the Rayleigh probability density function and is sketched in Figure 2-10 for two different values of σ^2.   Note that the most probable

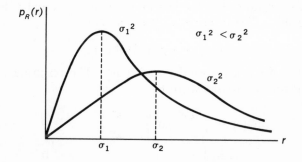

Figure 2-10   The Rayleigh probability density function.

value of the random variable is σ, but that the density function is not symmetrical about this maximum point.

The mean value of the Rayleigh-distributed random variable is easily computed from

$$\bar{R} = \int_0^\infty r p_R(r) \, dr = \int_0^\infty \frac{r^2}{\sigma^2} \exp\left[-\frac{r^2}{2\sigma^2}\right] dr$$

$$= \sqrt{\frac{\pi}{2}} \sigma$$

and the mean-square value from

$$\overline{R^2} = \int_0^\infty r^2 p_R(r) \, dr = \int_0^\infty \frac{r^3}{\sigma^2} \exp\left[-\frac{r^2}{2\sigma^2}\right] dr$$

$$= 2\sigma^2$$

The variance of R is therefore given by

$$\sigma_R{}^2 = \overline{R^2} - (\bar{R})^2 = \left(2 - \frac{\pi}{2}\right)\sigma^2 = 0.429\sigma^2$$

Note that this variance is *not* the same as the variance σ^2 of the Gaussian random variables that generate the Rayleigh random variable. It may also be noted that, unlike the Gaussian density function, both the mean and variance depend upon a single parameter (σ^2) and cannot be adjusted independently.

The use of the Rayleigh distribution in dealing with the envelopes of Gaussian random processes is treated in a later chapter.

Maxwell distribution.   A classical problem in thermodynamics is that of determining the probability density function of the velocity of a molecule in a perfect gas. The basic assumption is that each component of velocity is Gaussian with zero mean and a variance of $\sigma^2 = kT/m$, where k is Boltzmann's constant, T is the absolute temperature, and m is the mass of the molecule. The total velocity is, therefore,

$$V = \sqrt{V_x{}^2 + V_y{}^2 + V_z{}^2}$$

and is said to have a *Maxwell distribution*. The resulting probability density function can be shown to be

$$p_V(v) = \sqrt{\frac{2}{\pi}} \frac{v^2}{\sigma^3} \exp\left[-\frac{v^2}{2\sigma^2}\right] \qquad v \geq 0$$

$$= 0 \qquad\qquad\qquad\qquad v < 0$$

(2-22)

The mean value of a Maxwellian-distributed random variable (the average molecule velocity) can be found in the usual way and is

$$\bar{V} = \sqrt{\frac{8}{\pi}} \sigma$$

The mean-square value and variance can be shown to be

$$\overline{V^2} = 3\sigma^2$$

$$\sigma_V{}^2 = \overline{V^2} - (\bar{V})^2 = \left(3 - \frac{8}{\pi}\right)\sigma^2$$

$$= 0.453\sigma^2$$

The mean kinetic energy can be obtained from $\overline{V^2}$ since

$$\epsilon = \frac{1}{2}mV^2$$

and

$$E[\epsilon] = \frac{1}{2}m\overline{V^2} = \frac{3}{2}m\sigma^2 = \frac{3}{2}m\left(\frac{kT}{m}\right) = \frac{3}{2}kT$$

which is the classical result.

Chi-square distribution. A generalization of the above results arises if one defines a random variable as

$$X^2 = Y_1{}^2 + Y_2{}^2 + \cdots + Y_n{}^2$$

where $Y_1, Y_2, \ldots, Y_n$ are independent Gaussian random variables with 0 mean and variance 1. The random variable X^2 is said to have a *Chi-square distribution with n degrees of freedom* and the probability density function is

$$p(x^2) = \frac{(x^2)^{n/2-1}}{2^{n/2}(n/2 - 1)!} \exp\left[-\frac{x^2}{2}\right] \qquad x^2 \geq 0$$
$$= 0 \qquad\qquad\qquad\qquad\qquad x^2 < 0$$

(2-23)

With suitable normalization of random variables (so as to obtain unit variance), the power distribution discussed above is seen to be chi-square with $n = 1$. Likewise, in the Rayleigh distribution, the *square* of the miss-distance (R^2) is chi-square with $n = 2$; and in the Maxwell distribution, the square of the velocity (V^2) is chi-square with $n = 3$. This latter case would lead to the probability density function of molecule *energies*.

The mean and variance of a chi-square random variable are particularly simple because of the initial assumption of unit variance for the components. Thus,

$$\overline{X^2} = n$$
$$(\sigma_{X^2})^2 = 2n$$

Log-normal distribution. A somewhat different relationship to the Gaussian distribution arises in the case of random variables that are *defined* as the logarithms of other random variables. For example, in communication systems the attenuation of the signal power in the transmission path is frequently expressed in units of *nepers*, and is calculated from

$$A = \ln\left(\frac{W_{\text{out}}}{W_{\text{in}}}\right) \quad \text{nepers}$$

where W_{in} and W_{out} are the input and output signal powers respectively. An experimentally observed fact is that the attenuation A is very often quite close to being a Gaussian random variable. The question that arises, therefore, concerns the probability density function of the power ratio.

In order to generalize this result somewhat, let two random variables be related by

$$Y = \ln X$$

or, equivalently, by

$$X = e^Y$$

and assume that Y is Gaussian with a mean of $\bar{Y}$ and a variance $\sigma_Y{}^2$. By using (2-3) it is easy to show that the probability density function of X is

$$
\begin{aligned}
p_X(x) &= \frac{1}{\sqrt{2\pi}\,\sigma_Y x}\exp\left[-\frac{(\ln x - \bar{Y})^2}{2\sigma_Y{}^2}\right] \quad & x \geq 0 \\
&= 0 & x < 0
\end{aligned}
\tag{2-24}
$$

This is the *log-normal* probability density function. In engineering work base 10 is frequently used for the logarithm rather than base e, but it is simple to convert from one to the other. Some typical density functions are sketched in Figure 2-11.

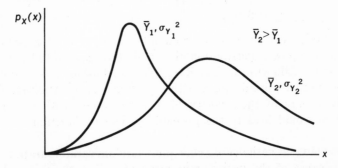

Figure 2-11   The log-normal probability density function.

The mean and variance of the log-normal random variable can be evaluated in the usual manner and become

$$\bar{X} = \exp\left[\bar{Y} + \frac{1}{2}\sigma_Y{}^2\right]$$

$$\sigma_X{}^2 = [\exp(\sigma_Y{}^2) - 1]\exp 2\left(\bar{Y} + \frac{1}{2}\sigma_Y{}^2\right)$$

2-7   Other Probability Density Functions

In addition to the density functions that are related to the Gaussian, there are many others that frequently arise in engineering. Some of these will be described here and an attempt will be made to discuss briefly the situations in which they arise.

Uniform distribution. The uniform distribution was mentioned in an earlier section and used for illustrative purposes; it will be generalized here. The uniform distribution usually arises in physical situations in which there is no preferred value for the random variable. For example, events that occur at random instants of time (such as the emission of radioactive particles) are often assumed to occur at times that are equally probable. The unknown phase angle associated with a sinusoidal source is usually assumed to be uniformly distributed over a range of 2π radians. The time position of pulses in a periodic sequence of pulses (such as a radar transmission) may be assumed to be uniformly distributed over an interval of one period, when the actual time position with respect to zero time is unknown. All of these situations will be employed in future examples.

The uniform probability density function may be represented generally as

$$p(x) = \frac{1}{x_2 - x_1} \qquad x_1 < x \leq x_2$$
$$= 0 \qquad\qquad \text{otherwise} \tag{2-25}$$

It is quite straightforward to show that

$$\bar{X} = \frac{1}{2}(x_1 + x_2) \tag{2-26}$$

and

$$\sigma_X{}^2 = \frac{1}{12}(x_2 - x_1)^2 \tag{2-27}$$

One of the important applications of the uniform distribution is in describing the errors associated with analog-to-digital conversion. This operation takes a continuous signal that can have any value at a given time instant and converts it into a binary number having a fixed number of binary digits. Since a fixed number of binary digits can represent only a discrete set of values, the difference between the actual value and the closest discrete value represents the error. This is illustrated in Figure 2-12. In order to determine the mean-square value of the error, it is *assumed* that the error is uniformly distributed over an interval from $-\Delta x/2$ to $\Delta x/2$ where Δx is the difference between the two closest levels. Thus, from (2-26), the mean error is zero, and from (2-27) the variance or mean-square error is $\frac{1}{12}(\Delta x)^2$.

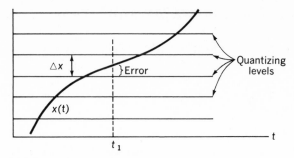

Figure 2-12 Error in analog-to-digital conversion.

Exponential and related distributions. It was noted in the discussion of the uniform distribution that events occurring at random time instants are often *assumed* to occur at times that are equally probable. Thus, if the average time interval between events is denoted $\bar{\tau}$, then the probability that an event will occur in a time interval Δt that is short compared to $\bar{\tau}$ is just $\Delta t/\bar{\tau}$ regardless of where that time interval is. From this assumption it is possible to derive the probability distribution function (and, hence, the density function) for the time interval between events.

In order to carry out this derivation, consider the sketch in Figure 2-13. It is assumed that an event has occurred at time t_0, and it is

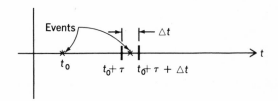

Figure 2-13 Time interval between events.

desired to determine the probability that the next event will occur at a random time lying between $t_0 + \tau$ and $t_0 + \tau + \Delta t$. If the distribution function for τ is $P(\tau)$, then this probability is just $P(\tau + \Delta t) - P(\tau)$. But the probability that the event occurred in the Δt interval must also be equal to the product of the probabilities of the independent events that the event did *not* occur between t_0 and $t_0 + \tau$ and the event that it did occur between $t_0 + \tau$ and $t_0 + \tau + \Delta t$. Since

$$1 - P(\tau) = \text{probability that event did } not \text{ occur between } t_0 \text{ and } t_0 + \tau$$

$$\frac{\Delta t}{\bar{\tau}} = \text{probability that it } did \text{ occur in } \Delta t$$

it follows that

$$P(\tau + \Delta t) - P(\tau) = [1 - P(\tau)]\left(\frac{\Delta t}{\bar{\tau}}\right)$$

Upon dividing both sides by Δt and letting Δt approach zero, it is clear that

$$\lim_{\Delta t \to 0} \frac{P(\tau + \Delta t) - P(\tau)}{\Delta t} = \frac{dP(\tau)}{d\tau} = \frac{1}{\bar{\tau}}[1 - P(\tau)]$$

The latter two terms comprise a first-order differential equation that can be solved to yield

$$P(\tau) = 1 - \exp\left[\frac{-\tau}{\bar{\tau}}\right] \qquad \tau \geq 0 \qquad\qquad \textbf{(2-28)}$$

In evaluating the arbitrary constant, use is made of the fact that $P(0) = 0$ since τ can never be negative.

The probability density function for the time interval between events can be obtained from (2-28) by differentiation. Thus,

$$p(\tau) = \frac{1}{\bar{\tau}} \exp\left[\frac{-\tau}{\bar{\tau}}\right] \qquad \tau \geq 0 \qquad\qquad \textbf{(2-29)}$$
$$= 0 \qquad\qquad\qquad \tau < 0$$

This is known as the exponential probability density function and is sketched in Figure 2-14 for two different values of average time interval.

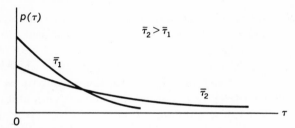

Figure 2-14 The exponential probability density function.

As would be expected, the mean value of τ is just $\bar{\tau}$. That is,

$$E[\tau] = \int_0^\infty \frac{\tau}{\bar{\tau}} \exp\left[\frac{-\tau}{\bar{\tau}}\right] d\tau = \bar{\tau}$$

The variance turns out to be

$$\sigma_\tau^2 = (\bar{\tau})^2$$

It may be noted that this density function (like the Rayleigh) is a single-parameter one. Thus the mean and variance are uniquely related and one determines the other.

As an illustration of the application of the exponential distribution, suppose that component failures in a spacecraft occur independently and uniformly with an average time between failures of 100 days. The spacecraft starts out on a 200-day mission with all components functioning.

What is the probability that it will complete the mission without a component failure? This is equivalent to asking for the probability that the time to the first failure is *greater* than 200 days; this is simply $[1 - P(200)]$ since $P(200)$ is the probability that this interval is *less* than (or equal to) 200 days. Hence, from (2-28)

$$1 - P(\tau) = 1 - \left[1 - \exp\left\{ \frac{-\tau}{\bar{\tau}} \right\} \right] = \exp\left\{ \frac{-\tau}{\bar{\tau}} \right\}$$

and for $\bar{\tau} = 100$, $\tau = 200$, this becomes

$$1 - P(200) = \exp\left[\frac{-200}{100} \right] = 0.1352$$

The random variable in the exponential distribution is the time interval between adjacent events. This can be generalized to make the random variable the time interval between any event and the kth following event. The probability distribution for this random variable is known as the *Erlang distribution* and the probability density function is

$$p_k(\tau) = \frac{\tau^{k-1} \exp\left[-\tau/\bar{\tau} \right]}{(\bar{\tau})^k (k-1)!} \qquad \tau \geq 0,\ k = 1, 2, 3, \ldots \tag{2-30}$$
$$= 0 \qquad\qquad\qquad \tau < 0$$

Such a random variable is said to be an *Erlang random variable of order k*. Note that the exponential distribution is simply the special case for $k = 1$. The mean and variance in the general case are $k\bar{\tau}$ and $k(\bar{\tau})^2$ respectively. The general Erlang distribution has a great many applications in engineering pertaining to the reliability of systems, the waiting times for users of a system (such as a telephone system or traffic system), and the number of channels required in a communication system to provide for a given number of users with random calling times and message lengths.

The Erlang distribution is also related to the *gamma distribution* by a simple change in notation. Letting $\beta = 1/\bar{\tau}$ and α be a continuous parameter that equals k for integral values, the Gamma distribution can be written as

$$p(\tau) = \frac{\beta^\alpha \tau^{\alpha-1}}{\Gamma(\alpha)} \exp\left[-\beta\tau \right] \qquad \tau \geq 0 \tag{2-31}$$
$$= 0 \qquad\qquad\qquad \tau < 0$$

The mean and variance of the Gamma distribution are α/β and α/β^2 respectively.

Delta distributions. It was noted earlier that when the possible events could assume only a discrete set of values, the appropriate probability density function consisted of a set of delta functions. It is desirable to formalize this concept somewhat and indicate some possible applications. As an example, consider the binary waveform illustrated in Figure 2-15.

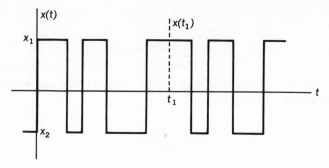

Figure 2-15   A general binary waveform.

Such a waveform arises in many types of communication systems or control systems since it obviously is the waveform with the greatest average power for a given peak value. It will be considered in more detail throughout the study of random processes, but the present interest is in a single random variable, $X = x(t_1)$, at a specified time instant. This random variable can assume only two possible values, x_1 or x_2; it is specified that it take on value x_1 with probability p_1 and value x_2 with probability $p_2 = 1 - p_1$.   Thus, the probability density function for X is

$$p(x) = p_1 \, \delta(x - x_1) + p_2 \, \delta(x - x_2) \tag{2-32}$$

The mean value associated with this random variable is evaluated easily as

$$\bar{X} = \int_{-\infty}^{\infty} x[p_1 \, \delta(x - x_1) + p_2 \, \delta(x - x_2)] \, dx$$
$$= p_1 x_1 + p_2 x_2$$

The mean-square value is determined similarly from

$$\overline{X^2} = \int_{-\infty}^{\infty} x^2[p_1 \, \delta(x - x_1) + p_2 \, \delta(x - x_2)] \, dx$$
$$= p_1 x_1{}^2 + p_2 x_2{}^2$$

Hence, the variance is

$$\sigma_X{}^2 = \overline{X^2} - (\bar{X})^2 = p_1 x_1{}^2 + p_2 x_2{}^2 - (p_1 x_1 + p_2 x_2)^2$$
$$= p_1 p_2 (x_1 - x_2)^2$$

in which use has been made of the fact that $p_2 = 1 - p_1$ in order to arrive at the final form.

It should be clear that similar delta distributions exist for random variables that can assume any number of discrete levels.   Thus, if there are n possible levels designated as $x_1, x_2, \ldots, x_n$, and the corresponding probabilities for each level are $p_1, p_2, \ldots, p_n$, then the probability density function is

$$p(x) = \sum_{i=1}^{n} p_i \, \delta(x - x_i) \tag{2-33}$$

in which

$$\sum_{i=1}^{n} p_i = 1$$

By using exactly the same techniques as above, the mean value of this random variable is shown to be

$$\bar{X} = \sum_{i=1}^{n} p_i x_i$$

and the mean-square value is

$$\overline{X^2} = \sum_{i=1}^{n} p_i x_i{}^2$$

From these, the variance becomes

$$\sigma_X{}^2 = \sum_{i=1}^{n} p_i x_i{}^2 - \left(\sum_{i=1}^{n} p_i x_i\right)^2$$

$$= \frac{1}{2} \sum_{i=1}^{n} \sum_{j=1}^{n} p_i p_j (x_i - x_j)^2$$

The multilevel delta distributions also arise in connection with communication and control systems, and in systems requiring analog-to-digital conversion. Typically the number of levels is an integer power of 2, so that they can be efficiently represented by a set of binary digits.

2-8   Conditional Probability Distribution and Density Functions

The concept of conditional probability was introduced in Section 1-6 in connection with the occurrence of discrete events. In that context it was the quantity expressing the probability of one event given that another event, *in the same probability space*, had already taken place. It is desirable to extend this concept to the case of continuous random variables. The discussion in the present section will be limited to definitions and examples involving a single random variable. The case of two or more random variables is considered in Chapter 3.

The first step is to define the conditional probability distribution function for a random variable X given that an event M has taken place. For the moment the event M is left arbitrary. The distribution function is denoted and defined by

$$P(x|M) = \Pr[X \leq x|M]$$
$$= \frac{\Pr\{X \leq x, M\}}{\Pr(M)} \qquad \Pr(M) > 0 \qquad \text{(2-34)}$$

where $\{X \leq x, M\}$ is the event of all outcomes ξ such that

$$X(\xi) \leq x \qquad \text{and} \qquad \xi \in M$$

where $X(\xi)$ is the value of the random variable X when the outcome of the experiment is ξ. Hence $\{X \leq x, M\}$ is the continuous counterpart of the set product used in the previous definition of (1-15). It can be shown that $P(x|M)$ is a valid probability distribution function and, hence, must have the same properties as any other distribution function. In particular, it has the following characteristics:

1. $0 \leq P(x|M) \leq 1 \qquad -\infty < x < \infty$
2. $P(-\infty|M) = 0 \qquad P(\infty|M) = 1$
3. $P(x|M)$ is *nondecreasing* as x increases
4. $\text{Pr}\,[x_1 < X \leq x_2|M] = P(x_2|M) - P(x_1|M)$
$$\geq 0 \qquad \text{for } x_1 < x_2$$

Now it is necessary to say something about the event M upon which the probability is conditioned. There are several different possibilities that arise. For example:

1. Event M may be an event that can be expressed in terms of the random variable X. Examples of this are considered in this section.
2. Event M may be an event that depends upon some other random variable, which may be either continuous or discrete. Examples of this are considered in Chapter 3.
3. Event M may be an event that depends upon both the random variable X and some other random variable. This is a more complicated situation which will not be considered at all.

As an illustration of the first possibility above, let M be the event

$$M = \{X \leq m\}$$

Then the conditional distribution function is, from (2-34),

$$P(x|M) = \text{Pr}\,\{X \leq x | X \leq m\} = \frac{\text{Pr}\,\{X \leq x, X \leq m\}}{\text{Pr}\,\{X \leq m\}}$$

There are now two possible situations—depending upon whether x or m is larger. If $x \geq m$, then the event that $X \leq m$ is contained in the event that $X \leq x$ and

$$\text{Pr}\,\{X \leq x, X \leq m\} = \text{Pr}\,\{X \leq m\}$$

Thus,

$$P(x|M) = \frac{\text{Pr}\,\{X \leq m\}}{\text{Pr}\,\{X \leq m\}} = 1 \qquad x \geq m$$

On the other hand, if $x \leq m$, then $\{X \leq x\}$ is contained in $\{X \leq m\}$ and

$$P(x|M) = \frac{\Pr\{X \leq x\}}{\Pr\{X \leq m\}} = \frac{P(x)}{P(m)}$$

The resulting conditional distribution function is shown in Figure 2-16.

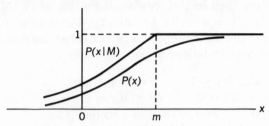

Figure 2-16   A conditional probability distribution function.

The conditional probability density function is related to the distribution function in the same way as before.   That is, when the derivative exists,

$$p(x|M) = \frac{dP(x|M)}{dx} \qquad \text{(2-35)}$$

This also has all the properties of a usual probability density function. That is,

1. $p(x|M) \geq 0 \qquad -\infty < x < \infty$
2. $\int_{-\infty}^{\infty} p(x|M)\, dx = 1$
3. $P(x|M) = \int_{-\infty}^{x} p(u|M)\, du$
4. $\int_{x_1}^{x_2} p(x|M)\, dx = \Pr[x_1 < X \leq x_2|M]$

If the example of Figure 2-16 is continued, the conditional probability density function is

$$p(x|M) = \frac{1}{P(m)} \frac{dP(x)}{dx} = \frac{p(x)}{P(m)} = \frac{p(x)}{\int_{-\infty}^{m} p(x)\, dx} \qquad x < m$$
$$= 0 \qquad\qquad\qquad\qquad\qquad\qquad\qquad x \geq m$$

This is sketched in Figure 2-17.

The conditional probability density function can also be used to find conditional means and conditional expectations.   For example, the conditional mean is

$$E[X|M] = \int_{-\infty}^{\infty} x p(x|M)\, dx \qquad \text{(2-36)}$$

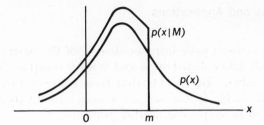

Figure 2-17 Conditional probability density function corresponding to Figure 2-16.

More generally, the conditional expectation of any $f(X)$ is

$$E[f(X)|M] = \int_{-\infty}^{\infty} f(x)p(x|M)\,dx \qquad (2\text{-}37)$$

As an illustration of the conditional mean, let the $p(x)$ in the above example be Gaussian so that

$$p(x) = \frac{1}{\sqrt{2\pi}\,\sigma} \exp\left[-\frac{(x-\bar{X})^2}{2\sigma^2}\right]$$

In order to make the example simple, let $m = \bar{X}$ so that

$$P(m) = \int_{-\infty}^{m=\bar{X}} \frac{1}{\sqrt{2\pi}\,\sigma} \exp\left[-\frac{(x-\bar{X})^2}{2\sigma^2}\right] dx = \frac{1}{2}$$

Thus,

$$p(x|M) = \frac{p(x)}{\frac{1}{2}} = \frac{2}{\sqrt{2\pi}\,\sigma} \exp\left[-\frac{(x-\bar{X})^2}{2\sigma^2}\right] \qquad x < \bar{X}$$
$$= 0 \qquad x \geq \bar{X}$$

Hence, the conditional mean is

$$E[x|M] = \int_{-\infty}^{\bar{X}} \frac{2x}{\sqrt{2\pi}\,\sigma} \exp\left[\frac{(x-\bar{X})^2}{2\sigma^2}\right] dx$$
$$= \int_{-\infty}^{0} \frac{2(u+\bar{X})}{\sqrt{2\pi}\,\sigma} \exp\left[-\frac{u^2}{2\sigma^2}\right] du$$
$$= \bar{X} - \sqrt{\frac{2}{\pi}}\,\sigma$$

In words, this result says that the expected value or conditional mean of a Gaussian random variable, given that the random variable is less than its mean, is just

$$\bar{X} - \sqrt{\frac{2}{\pi}}\,\sigma$$

2-9 Examples and Applications

The preceding sections have introduced some of the basic concepts concerning the probability distribution and density functions for a continuous random variable. Before extending these concepts to more than one variable, it is desirable to consider a few examples illustrating how they might be applied to simple engineering problems.

As a first example, consider the elementary voltage-regulating circuit shown in Figure 2-18(a). It employs a Zener diode having an idealized

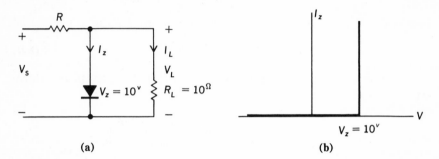

(a)   (b)

Figure 2-18 Zener diode voltage regulator: (a) voltage-regulating circuit and (b) Zener diode characteristic.

current-voltage characteristic as shown in Figure 2-18(b). Note that current is zero until the voltage reaches the breakdown value ($V_z = 10$) and from then on is limited by the external circuit, while the voltage across the diode remains constant. Such a circuit is often used to limit the voltage applied to solid-state devices. For example, the R_L indicated in the circuit may be a transistorized amplifier designed to work at 9 V and that is damaged if the voltage exceeds 10 V. The supply voltage, V_s, is from a power supply whose nominal voltage is 12 V, but whose actual voltage contains a sawtooth ripple and, hence, is a random variable. For purposes of this example, it will be assumed that this random variable has a uniform distribution over the interval from 9 to 15 V.

Zener diodes are rated in terms of their ability to dissipate power as well as their breakdown voltage. It will be assumed that the average power rating of this diode is $W_z = 3$ W. It is then desired to find the value of series resistance, R, needed to limit the mean dissipation in the Zener diode to this rated value.

When the Zener diode is conducting, the voltage across it is $V_z = 10$, and the current through it is

$$I_z = \frac{V_s - V_z}{R} - I_L \quad \text{and} \quad V_s > \frac{V_z(R + R_L)}{R_L} = \frac{10(R + 10)}{10}$$

where the load current, I_L, is 1 A. The power dissipated in the diode is

$$W_z = V_z I_z = \frac{V_z(V_s - V_z)}{R} - I_L V_z$$

$$= \frac{10 V_s - 100}{R} - 10 \qquad V_s > R + 10$$

A sketch of this power as a function of the supply voltage V_s is shown in Figure 2-19, and the probability density functions of V_s and W_z are

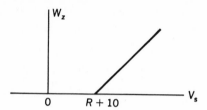

Figure 2-19 Relation between diode power dissipation and supply voltage.

shown in Figure 2-20. Note that the density function of W_z has a large delta function at zero, since the diode is not conducting most of the time,

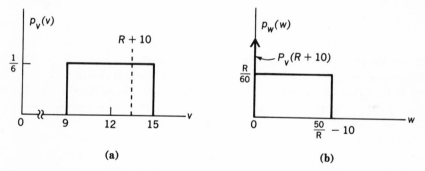

(a)   (b)

Figure 2-20 Probability density functions for supply voltage and diode power dissipation: (a) probability density function for V_s and (b) probability density function for W_z.

but is uniform for larger values of W since W_z and V_s are linearly related in this range. From the previous discussion of transformations of density functions in Section 2-3, it is easy to show that

$$p_W(w) = P_V(R + 10)\, \delta(w) + \frac{R}{10} p_V\left(\frac{Rw}{10} + R + 10\right) \qquad 0 \le w \le \frac{50}{R} - 10$$

$$= 0 \qquad\qquad\qquad\qquad\qquad\qquad \text{elsewhere}$$

where $P_V(\cdot)$ is the distribution function of V_s. Hence, the area of the delta function is simply the probability that the supply voltage V_s is *less* than the value that causes diode conduction to start.

The mean value of diode power dissipation is now given by

$$E[W_z] = \bar{W}_z = \int_{-\infty}^{\infty} w p_W(w) \, dw$$

$$= \int_{-\infty}^{\infty} w P_V(R + 10) \, \delta(w) \, dw$$

$$+ \int_{0}^{\infty} w \left(\frac{R}{10}\right) p_V \left(\frac{Rw}{10} + R + 10\right) dw$$

The first integral has a value of zero (since the delta function is at $w = 0$) and the second integral can be written in terms of the uniform density function $[p_V(v) = \frac{1}{6}, \, 9 < v \leq 15]$, as

$$\bar{W}_z = \int_{0}^{(50/R) - 10} w \left(\frac{R}{10}\right) \left(\frac{1}{6}\right) dw = \frac{(5 - R)^2}{1.2R}$$

Since the mean value of diode power dissipation is to be less than or equal to 3 watts, it follows that

$$\frac{(5 - R)^2}{1.2R} \leq 3 \qquad 0 < R \leq 5$$

from which

$$R \geq 2.19 \, \Omega$$

It may now be concluded that any value of R greater than 2.19 Ω would be satisfactory from the standpoint of limiting the mean value of power dissipation in the Zener diode to 3 watts. The actual choice of R would be determined by the desired value of output voltage at the nominal supply voltage of 12 V. If this desired voltage is 9 V (as suggested above) then R must be

$$R = \frac{3}{\frac{9}{10}} = 3.33 \, \Omega$$

which is greater than the minimum value of 2.19 Ω and, hence, would be satisfactory.

As another example, consider the problem of selecting a multiplier resistor for a dc voltmeter as shown in Figure 2-21. It will be assumed that the dc instrument produces full-scale deflection when 100 μA is

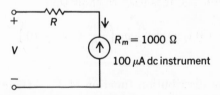

Figure 2-21   Selection of a voltmeter resistor.

passing through the coil and has a resistance of 1000 Ω. It is desired to select a multiplier resistor R such that this instrument will read full scale when 10 V is applied. Thus, the nominal value of R to accomplish this (which will be designated as R^*) is

$$R^* = \frac{10}{10^{-4}} - 1000 = 9.9 \times 10^4 \ \Omega$$

However, the actual resistor used will be selected at random from a bin of resistors marked 10^5 Ω. Because of manufacturing tolerances, the actual resistance is a random variable having a mean of 10^5 and a standard deviation of 1000 Ω. It will also be *assumed* that the actual resistance is a Gaussian random variable. (This is a customary assumption when deviations around the mean are small, even though it can never be precisely true for quantities that must be always positive, like the resistance.) On the basis of these assumptions it is desired to find the probability that the resulting voltmeter will be accurate to within 2 percent.[6]

The smallest value of resistance that would be acceptable is

$$R_{\min} = \frac{10 - 0.2}{10^{-4}} - 1000 = 9.7 \times 10^4$$

while the largest value is

$$R_{\max} = \frac{10 + 0.2}{10^{-4}} - 1000 = 10.1 \times 10^4$$

The probability that a resistor selected at random will fall between these two limits is

$$P_c = \Pr\left[9.7 \times 10^4 < R \leq 10.1 \times 10^4\right] = \int_{9.7 \times 10^4}^{10.1 \times 10^4} p_R(r) \, dr \quad \textbf{(2-38)}$$

where $p_R(r)$ is the Gaussian probability density function for R and is given by

$$p_R(r) = \frac{1}{\sqrt{2\pi}\,(1000)} \exp\left[-\frac{(r - 10^5)^2}{2(10^6)}\right]$$

The integral in (2-38) can be expressed in terms of the standard normal distribution function, $\Phi(\cdot)$, as discussed in Section 2-5. Thus, P_c becomes

$$P_c = \Phi\left(\frac{10.1 \times 10^4 - 10^5}{10^3}\right) - \Phi\left(\frac{9.7 \times 10^4 - 10^5}{10^3}\right)$$

---

[6] This is interpreted to mean that the error in voltmeter reading due to the resistor value is less than or equal to 2 percent of the full scale reading.

which can be simplified to

$$P_c = \Phi(1) - \Phi(-3)$$
$$= \Phi(1) - [1 - \Phi(3)]$$

Using the tables in Appendix D, this becomes

$$P_c = 0.8413 - [1 - 0.9987] = 0.8400$$

Thus, it appears that even though the resistors are selected from a supply that is nominally incorrect, there is still a substantial probability that the resulting instrument will be within acceptable limits of accuracy.

The third and final example will consider an application of conditional probability. This example considers a traffic measurement system that is measuring the speed of all vehicles on an expressway and recording those speeds in excess of the speed limit of 70 miles per hour (mph). If the vehicle speed is a random variable with a Rayleigh distribution and a most probable value equal to 50 mph, it is desired to find the mean value of the excess speed. This is equivalent to finding the conditional mean of vehicle speed, given that the speed is greater than the limit, and subtracting the limit from it.

Letting the vehicle speed be S, the conditional distribution function that is sought is

$$P[s|S > 70] = \frac{\Pr\{S \le s, S > 70\}}{\Pr\{S > 70\}} \qquad \text{(2-39)}$$

Since the numerator is nonzero only when $s > 70$, (2-39) can be written as

$$P[s|S > 70] = 0 \qquad\qquad s \le 70$$
$$= \frac{P(s) - P(70)}{1 - P(70)} \qquad s > 70 \qquad \text{(2-40)}$$

where $P(\cdot)$ is the probability distribution function for the random variable S. The numerator of (2-40) is simply the probability that S is between 70 and s, while the denominator is the probability that S is greater than 70.

The conditional probability density function is found by differentiating (2-40) with respect to s. Thus,

$$p(s|S > 70) = 0 \qquad\qquad s \le 70$$
$$= \frac{p(s)}{1 - P(70)} \qquad s > 70$$

where $p(s)$ is the Rayleigh density function given by

$$p(s) = \frac{s}{(50)^2} \exp\left[-\frac{s^2}{2(50)^2}\right] \qquad s \ge 0 \qquad \text{(2-41)}$$
$$= 0 \qquad\qquad\qquad s < 0$$

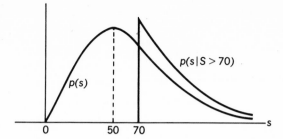

Figure 2-22 Conditional and unconditional density functions for a Rayleigh-distributed random variable.

These functions are sketched in Figure 2-22.

The quantity $P(70)$ is easily obtained from (2-41) as

$$P(70) = \int_0^{70} \frac{s}{(50)^2} \exp\left[-\frac{s^2}{2(50)^2}\right] ds = 1 - \exp\left[-\frac{49}{50}\right]$$

Hence,

$$1 - P(70) = \exp\left[-\frac{49}{50}\right]$$

The conditional expectation is given by

$$E[S|S > 70] = \frac{1}{\exp\left[-49/50\right]} \int_{70}^{\infty} \frac{s^2}{(50)^2} \exp -\left[\frac{s^2}{2(50)^2}\right] ds$$

$$= 70 + 50 \sqrt{2\pi} \exp\left[\frac{49}{50}\right] \left\{1 - \Phi\left(\frac{7}{5}\right)\right\}$$

$$= 70 + 27.2$$

Thus, the mean value of the excess speed is 27.2 miles per hour. Although it is clear from this result that the Rayleigh model is not a realistic one for traffic systems (since 27.2 miles per hour excess speed is much too large for the actual situation), the above example does illustrate the general technique for finding conditional means.

■ PROBLEMS

2-1 When two dice are tossed the event of interest is the sum of the uppermost faces. Let this sum be the random variable.

a. Plot the distribution function of the random variable.
b. What is the probability that the random variable is between 7 and 9, inclusive?
c. If five dice are tossed simultaneously, plot the distribution function of the number of ones that occur.

2-2   A probability distribution function for a random variable X has the form shown

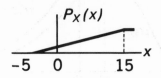

a. What is $P_X(0)$?
b. What is the probability the random variable X lies in the interval $0 < X \leq \infty$?
c. What is the probability that X lies in the interval $5 \leq X \leq 10$?

2-3   A power supply has five intermittent loads connected to it and each load, when in operation, draws 2 W.   Each load is in operation one-quarter of the time and independently of all other loads.

a. Plot the distribution function of the power delivered by the supply.
b. Plot the density function of the power delivered by the supply.
c. If the power supply can deliver only 6 W, what is the probability that the load requirement will be met?

2-4   A random variable X has a probability density function given by

$$p(x) = A\epsilon^{-x} \qquad x > 1$$
$$= 0 \qquad x \leq 1$$

a. Determine the value of the constant A.
b. What is the probability that the random variable X is in the range $1 < X \leq 2$?
c. Determine and sketch the probability distribution function of X.

2-5   For the random variable in Problem 2-4 find:

a. The mean value $\bar{X}$.
b. The mean-square value $\overline{X^2}$.
c. The variance σ^2.

2-6   A random variable has a probability density function of

$$p(x) = \tfrac{1}{10}, \qquad -3 \leq x \leq 7$$
$$= 0, \qquad \text{elsewhere}$$

a. Find the mean value, $\bar{X}$.
b. Find the mean-square value, $\overline{X^2}$.

c. Find the variance, $\sigma_X{}^2$.
d. Find the fourth central moment, $E[(X - \bar{X})^4]$.

2-7   **a.** Prove that for a Gaussian distribution with mean $\bar{X}$, that

$$P(x - \bar{X}) = 1 - P(\bar{X} - x)$$

where $P(x)$ is the probability distribution function for a *zero mean* Gaussian random variable.
b. Suppose that X is a Gaussian random variable.   If 10 percent of the values of X are below 60 and 5 percent are above 90, what are the mean and variance of X?

2-8   A Gaussian random variable X has a mean value of 2 and a variance of 4.

a. Write the probability density function, $p(x)$.
b. What is the value of the density function at $x = 0$, $x = 2$, and $x = 4$?
c. What is the probability that the random variable X will have a value greater than 0?   Greater than 2?
d. What is the probability that the random variable X will lie in the range $(\bar{X} - \sigma) < X \leq (\bar{X} + \sigma)$?

2-9   A Gaussian random current has a mean value of 1 A and a standard deviation of 4 A.   It is flowing in a resistance of 10 Ω.

a. What is the mean value of power dissipated in the resistor?
b. What is the variance of the power dissipated in the resistor?
c. How does the mean value of the current affect the mean and variance of the power?

2-10   Show that the probability of the power associated with a Gaussian voltage or current being zero is zero even though the density function is infinite at that point as given by equation (2-17).

2-11   Marbles rolling on a flat surface have components of velocity in orthogonal directions that are independent Gaussian random variables with zero mean and a standard deviation of 10 feet per second.

a. What is the most probable speed (in other words, total velocity) of the marbles?
b. What is the mean value of speed?
c. What is the probability of finding a marble with a speed greater than 20 feet per second?

2-12   The probability density function shown is the *Simpson* or *triangular* density function.   The triangle is isosceles.

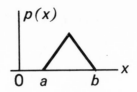

a. Write an expression for the probability density function, $p(x)$.
b. Determine the mean value.
c. Determine the variance.

2-13 A random variable, θ, is uniformly distributed over a range of 0 to 2π.   Another random variable, X, is related to θ by

$$X = \cos \theta$$

a. What is the probability density function of X?
b. What is the mean value of X?
c. What is the variance of X?

2-14 a. For the random variable X of problem 2-13, find the conditional probability density function $p(x|M)$, where M is the event $\{0 \le \theta \le \pi/2\}$. Sketch this density function.
b. Find the conditional mean, $E[X|M]$, for the same event M.

2-15 In a radar system the reflected signal pulses may have amplitudes that are Rayleigh distributed.   Let the probability density function of the pulse amplitudes be

$$p(r) = re^{-r^2/2}, \qquad r \ge 0$$
$$\quad\;\; = 0, \qquad\qquad r < 0$$

However, the only pulses that are displayed on the radar scope are those for which the pulse amplitude R is greater than some threshold, r_o, in order that the effects of system noise can be substantially eliminated.

a. Determine and sketch the probability density function of the displayed pulses, in other words, $p(r|R > r_o)$.
b. Determine the conditional mean for these displayed pulses.

2-16 An event M is given by $M = \{m_1 < X \le m_2\}$.

a. Show that the conditional probability density function, $P(x|M)$ is given by

$$P(x|M) = \frac{P(x) - P(m_1)}{P(m_2) - P(m_1)} \qquad m_1 < x \le m_2$$
$$\qquad\quad\; = 0 \qquad\qquad\qquad\;\; x \le m_1$$
$$\qquad\quad\; = 1 \qquad\qquad\qquad\;\; x > m_2$$

b. Given that

$$p(x) = \tfrac{1}{8}x \qquad 0 < x \le 4$$
$$= 0 \qquad \text{elsewhere}$$

find $p(x|M)$ where M is the event $M = \{1 < X \le 3\}$.

2-17 A random variable X is transformed linearly into another random variable Y by the relation

$$Y = aX + b$$

Such a linear transformation is a very common one in system analysis.

a. Find the probability density function $p_Y(y)$ in terms of the density function $p_X(x)$.
b. If X has a mean of $\bar{X}$ and a variance of $\sigma_X{}^2$, find the mean, $\bar{Y}$, and variance, $\sigma_Y{}^2$, of the random variable Y.

2-18 Different types of electronic ac voltmeters produce deflections that are proportional to different characteristics of the applied waveforms. In almost all cases, however, the scale is calibrated so that the voltmeter correctly indicates the rms value of a *sine wave*. For other types of waveforms the scale reading may not be equal to the rms value.

Suppose the following instruments are connected to a Gaussian random voltage having zero mean and a standard deviation of 10 V.   What will each read?

a. An instrument in which the deflection is proportional to the average of the full-wave rectified waveform.   That is, if $X(t)$ is applied, the deflection is proportional to $E[|X(t)|]$.
b. An instrument in which the deflection is proportional to the average of the envelope of the waveform.   Remember that the envelope of a Gaussian waveform is Rayleigh distributed.

■**REFERENCES**

See References for Chapter 1, particularly Beckmann, Lanning and Battin, Papoulis, and Parzen.

CHAPTER

3

Several Random Variables

3-1　Two Random Variables

All of the discussion so far has concentrated on situations involving a single random variable. This random variable may be, for example, the value of a voltage or current at a particular instant of time. It should be apparent, however, that saying something about a random voltage or current at only one instant of time is not adequate as a means of describing the nature of complete time functions. Such time functions, even if of finite duration, have an infinite number of random variables associated with them. This raises the question, therefore, of how one can extend the probabilistic description of a single random variable to include the more realistic situation of continuous time functions. The purpose of this section is to take the first step of that extension by considering *two* random variables. It might appear that this is an insignificant advance toward the goal of dealing with an infinite number of random variables, but it will become apparent later than this is really all that is needed, *provided that* the two random variables are separated in time by an arbitrary time interval. That is, if the random variables associated with *any* two instants of time can be described, then all of the information is available in order to carry out most of the usual types of systems analysis. Another situation that can arise in systems analysis is that in which it is desired to find the relation between the input and output of the system, either at the same instant of time or at two different time instants.   Again, only two random variables are involved.

In order to deal with situations involving two random variables it is necessary to extend the concepts of probability distribution and density functions that were discussed in the last chapter.   Let the two random variables be designated as X and Y and define a *joint probability distribution function* as

$$P(x,y) = \Pr [X \leq x, Y \leq y]$$

Note that this is simply the probability of the event that the random variable X is less than or equal to x *and* that the random variable Y is less than or equal to y.   As such, it is a straightforward extension of the probability distribution function for one random variable.

The joint probability distribution function has properties that are quite analogous to those discussed previously for a single variable. These may be summarized as follows:

1. $0 \leq P(x,y) \leq 1$ $-\infty < x < \infty$ $-\infty < y < \infty$
2. $P(-\infty,y) = P(x,-\infty) = P(-\infty,-\infty) = 0$
3. $P(\infty,\infty) = 1$
4. $P(x,y)$ is a nondecreasing function as either x or y, or both, increase
5. $P(\infty,y) = P_Y(y)$ $P(x,\infty) = P_X(x)$

In item 5 above, the subscripts on $P_Y(y)$ and $P_X(x)$ are introduced to indicate that these two distribution functions are not necessarily the same mathematical function of their respective arguments.

As an example of joint probability distribution functions consider the outcomes of tossing two coins.   Let X be a random variable associated with the first coin; let it have a value of 0 if a tail occurs and a value of 1 if a head occurs.   Similarly let Y be associated with the second coin and also have possible values of 0 and 1.   The joint distribution function, $P(x,y)$, is shown in Figure 3-1.   Note that it satisfies all of the properties listed above.

It is also possible to define a *joint probability density function* by differentiating the distribution function.   Since there are two independent variables, however, this differentiation must be done partially.   Thus,

$$p(x,y) = \frac{\partial^2 P(x,y)}{\partial x\, \partial y} \tag{3-1}$$

and the sequence of differentiation is immaterial.   The probability element is

$$p(x,y)\, dx\, dy = \Pr [x < X \leq x + dx, y < Y \leq y + dy] \tag{3-2}$$

The properties of the joint probability density function are quite

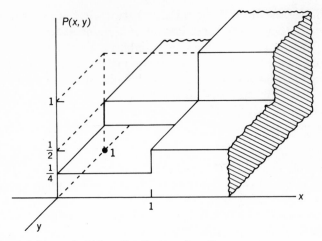

Figure 3-1 A joint probability distribution function.

analogous to those of a single random variable and may be summarized as follows:

1. $p(x,y) \geq 0$ $-\infty < x < \infty$ $-\infty < y < \infty$

2. $\int_{-\infty}^{\infty} \int_{-\infty}^{\infty} p(x,y) \, dx \, dy = 1$

3. $P(x,y) = \int_{-\infty}^{x} \int_{-\infty}^{y} p(u,v) \, dv \, du$

4. $p_X(x) = \int_{-\infty}^{\infty} p(x,y) \, dy$ $p_Y(y) = \int_{-\infty}^{\infty} p(x,y) \, dx$

5. $\Pr[x_1 < X \leq x_2, y_1 < Y \leq y_2] = \int_{x_1}^{x_2} \int_{y_1}^{y_2} p(x,y) \, dy \, dx$

Note that item 2 implies that the *volume* beneath any joint probability density function must be unity.

As a simple illustration of a joint probability density function, consider a pair of random variables having a density function that is constant between x_1 and x_2 and between y_1 and y_2. Thus,

$$p(x,y) = \frac{1}{(x_2 - x_1)(y_2 - y_1)} \quad \begin{cases} x_1 < x \leq x_2 \\ y_1 < y \leq y_2 \end{cases} \qquad \text{(3-3)}$$
$$= 0 \qquad\qquad\qquad\qquad \text{elsewhere}$$

This density function and the corresponding distribution function are shown in Figure 3-2.

The joint probability density function can be used to find the expected value of functions of two random variables in much the same way as with the single variable density function. In general, the expected value of any function, $f(X,Y)$, can be found from

$$E[f(X,Y)] = \int_{-\infty}^{\infty} \int_{-\infty}^{\infty} f(x,y) p(x,y) \, dx \, dy \qquad \text{(3-4)}$$

One such expected value that will be considered in great detail in a sub-

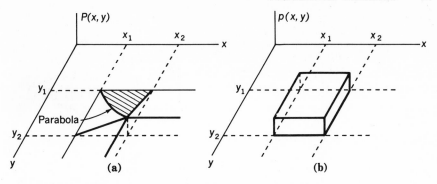

Figure 3-2 (a) Joint distribution and (b) density functions.

sequent section arises when $f(X,Y) = XY$. The expected value is known as the *correlation* and is given by

$$E[XY] = \int_{-\infty}^{\infty} \int_{-\infty}^{\infty} xyp(x,y) \, dx \, dy \tag{3-5}$$

As a simple example of the calculation, consider the joint density function shown in Figure 3-2(b). Since it is zero everywhere except in the specific region, (3-4) may be written as

$$E[XY] = \int_{x_1}^{x_2} dx \int_{y_1}^{y_2} xy \left[\frac{1}{(x_2 - x_1)(y_2 - y_1)} \right] dy$$

$$= \frac{1}{(x_2 - x_1)(y_2 - y_1)} \left[\frac{x^2}{2} \Big|_{x_1}^{x_2} \right] \left[\frac{y^2}{2} \Big|_{y_1}^{y_2} \right]$$

$$= \tfrac{1}{4}(x_1 + x_2)(y_1 + y_2)$$

Item 4 in the above list of properties of joint probability density functions indicates that the *marginal* probability density functions can be obtained by integrating the joint density over the other variable. Thus, for the density function in Figure 3-2(b), it follows that

$$p_X(x) = \int_{y_1}^{y_2} \frac{1}{(x_2 - x_1)(y_2 - y_1)} \, dy$$

$$= \frac{1}{(x_2 - x_1)(y_2 - y_1)} \left[y \Big|_{y_1}^{y_2} \right]$$

$$= \frac{1}{x_2 - x_1} \tag{3-6a}$$

and

$$p_Y(y) = \int_{x_1}^{x_2} \frac{1}{(x_2 - x_1)(y_2 - y_1)} \, dx$$

$$= \frac{1}{(x_2 - x_1)(y_2 - y_1)} \left[x \Big|_{x_1}^{x_2} \right]$$

$$= \frac{1}{y_2 - y_1} \tag{3-6b}$$

Exercise 3-1

In the example discussed above assume that $x_1 = 1$, $x_2 = 2$, $y_1 = 1$, and $y_2 = 4$. Find (a) the probability that $X \geq 1$ and $Y \geq 3$, (b) the probability that $X \geq 1.5$ and $Y \leq 2$ and (c) $E[XY]$.

Answer: $\frac{15}{4}$, $\frac{1}{3}$, $\frac{1}{6}$

3-2 Conditional Probability—Revisited

Now that the concept of joint probability for two random variables has been introduced, it is possible to extend the previous discussion of conditional probability. The previous definition of the conditional probability density function left the given event M somewhat arbitrary—although some specific examples were given. In the present discussion the event M will be related to another random variable, Y.

There are several different ways in which the given event M can be defined in terms of Y. For example, M might be the event $Y \leq y$ and, hence, Pr (M) would be just the marginal distribution function of Y, that is, $P_Y(y)$. From the basic definition of the conditional distribution function given in (2-34) of the previous chapter, it would follow that

$$P_X(x|Y \leq y) = \frac{\text{Pr } [X \leq x, M]}{\text{Pr } (M)} = \frac{P(x,y)}{P_Y(y)} \tag{3-7}$$

Another possible definition of M is that it is the event $y_1 < Y \leq y_2$. The definition of (2-34) now leads to

$$P_X(x|y_1 < Y \leq y_2) = \frac{P(x,y_2) - P(x,y_1)}{P_Y(y_2) - P_Y(y_1)} \tag{3-8}$$

In both of the above situations, the event M has a nonzero probability, that is, Pr $(M) > 0$. However, the most common form of conditional probability is one in which M is the event that $Y = y$; in almost all these cases Pr $(M) = 0$, since Y is continuously distributed. Since the conditional distribution function is defined as a ratio, it usually still exists even in these cases. It can be obtained from (3-8) by letting $y_1 = y$ and $y_2 = y + \Delta y$ and by taking a limit as Δy approaches zero. Thus,

$$P_X(x|Y = y) = \lim_{\Delta y \to 0} \frac{P(x, y + \Delta y) - P(x,y)}{P_Y(y + \Delta y) - P_Y(y)} = \frac{\partial P(x,y)/\partial y}{\partial P_Y(y)/\partial y}$$
$$= \frac{\int_{-\infty}^{x} p(u,y)\, du}{p_Y(y)} \tag{3-9}$$

The corresponding conditional density function is

$$p_X(x|Y = y) = \frac{\partial P_X(x|Y = y)}{\partial x} = \frac{p(x,y)}{p_Y(y)} \tag{3-10}$$

and this is the form that will be most commonly used. By interchanging X and Y it follows that

$$p_Y(y|X = x) = \frac{p(x,y)}{p_X(x)} \tag{3-11}$$

Because this form of conditional density function is so frequently used, it is convenient to adopt a shorter notation. Thus, when there is no danger of ambiguity, the conditional density functions will be written as

$$p(x|y) = \frac{p(x,y)}{p_Y(y)} \tag{3-12}$$

$$p(y|x) = \frac{p(x,y)}{p_X(x)} \tag{3-13}$$

From these two equations one can obtain the continuous version of *Bayes' theorem*, which was given by (1-19) for the discrete case. Thus, eliminating $p(x,y)$ leads directly to

$$p(y|x) = \frac{p(x|y)p_Y(y)}{p_X(x)} \tag{3-14}$$

It is also possible to obtain the total probability from (3-12) or (3-13) by noting that

$$p_X(x) = \int_{-\infty}^{\infty} p(x,y)\, dy = \int_{-\infty}^{\infty} p(x|y)p_Y(y)\, dy \tag{3-15}$$

and

$$p_Y(y) = \int_{-\infty}^{\infty} p(x,y)\, dx = \int_{-\infty}^{\infty} p(y|x)p_X(x)\, dx \tag{3-16}$$

These equations are the continuous counterpart of (1-18), which applied to the discrete case.

The use of conditional density functions arises in many different situations, but one of the most common (and probably the simplest) is that in which some observed quantity is the sum of two quantities—one of which is usually considered to be a signal while the other is considered to be a noise. Suppose, for example, that a signal $X(t)$ is perturbed by additive noise $N(t)$ and that the sum of these two, $Y(t)$, is the only quantity that can be observed. Hence, at some time instant, there are three random variables related by

$$Y = X + N$$

and it is desired to find the conditional probability density function of X given the observed value of Y, that is, $p(x|y)$. The reason for being interested in this is that the most probable values of X, given the observed value Y, may be a reasonable guess, or estimate, of the true value of X when X can only be observed in the presence of noise. From Bayes' theorem this conditional probability is

$$p(x|y) = \frac{p(y|x)p_X(x)}{p_Y(y)}$$

But if X is given, as implied by $p(y|x)$, then the only randomness about Y is the noise N, and it is assumed that its density function, $p_N(n)$, is known. Thus, since $N = Y - X$, and X is given,

$$p(y|x) = p_N(n = y - x) = p_N(y - x)$$

The desired conditional probability density, $p(x|y)$, can now be written as

$$p(x|y) = \frac{p_N(y - x)p_X(x)}{p_Y(y)} = \frac{p_N(y - x)p_X(x)}{\int_{-\infty}^{\infty} p_N(y - x)p_X(x)\,dx} \tag{3-17}$$

in which the integral in the denominator is obtained from (3-16).   Thus, if the a priori density function of the signal, $p_X(x)$, and the noise density function, $p_N(n)$, are known, it becomes possible to determine the conditional density function, $p(x|y)$.   When some particular value of Y is observed, say y_1, then the value of x for which $p(x|y_1)$ is a maximum is a good estimate for the true value of X.

As a specific example of the above application of conditional probability, suppose that the signal random variable, X, has an exponential density function so that

$$p_X(x) = b \exp(-bx) \qquad x \geq 0$$
$$= 0 \qquad\qquad\quad x < 0$$

Such a density function might arise, for example, as a signal from a space probe in which the *time intervals* between counts of high-energy particles are converted to *voltage amplitudes* for purposes of transmission back to earth.   The noise that is added to this signal is assumed to be Gaussian, with zero mean, so that its density function is

$$p_N(n) = \frac{1}{\sqrt{2\pi}\,\sigma_N} \exp\left[-\frac{n^2}{2\sigma_N{}^2}\right]$$

The marginal density function of Y, which appears in the denominator of (3-17), now becomes

$$p_Y(y) = \int_0^{\infty} \frac{b}{\sqrt{2\pi}\,\sigma_N} \exp\left[-\frac{(y - x)^2}{2\sigma_N{}^2}\right] \exp[-bx]\,dx$$

$$= \frac{b}{2} \exp\left[-by + \frac{b^2\sigma_N{}^2}{2}\right]\left[1 + \mathrm{erf}\left(\frac{y - b\sigma_N{}^2}{\sqrt{2}\,\sigma_N}\right)\right]^1 \tag{3-18}$$

---

[1] The *error function* is related to the normal probability distribution function and is defined as

$$\mathrm{erf}(z) = \frac{2}{\sqrt{\pi}} \int_0^z \epsilon^{-u^2}\,du = 2\phi(\sqrt{2}\,z) - 1$$

It should be noted, however, that if one is interested only in locating the maximum of $p(x|y)$, it is not necessary to evaluate $p_Y(y)$ since it is *not* a function of x. Hence, for a given Y, $p_Y(y)$ is simply a constant.

The desired conditional density function can now be written, from (3-17), as

$$p(x|y) = \frac{b}{\sqrt{2\pi}\ \sigma_N p_Y(y)} \exp\left[-\frac{(y-x)^2}{2\sigma_N^2}\right] \exp\left[-bx\right] \qquad x \geq 0$$
$$= 0 \qquad\qquad\qquad x < 0$$

This may also be written as

$$p(x|y) = \frac{b}{\sqrt{2\pi}\ \sigma_N p_Y(y)} \exp\left\{-\frac{1}{2\sigma_N^2}\left[x^2 - 2(y - b\sigma_N^2)x + y^2\right]\right\} \qquad x \geq 0$$
$$= 0 \qquad\qquad\qquad\qquad\qquad\qquad x < 0$$

$$(3\text{-}19)$$

and this is sketched in Figure 3-3 for two different values of y.

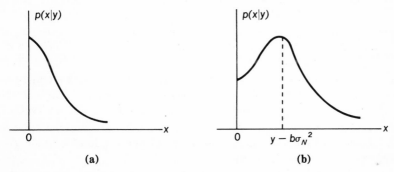

(a)　　　　　　　　　(b)

Figure 3-3　The conditional density function, $p(x|y)$: (a) case for $y < b\sigma_N^2$ and (b) case for $y > b\sigma_N^2$.

It was noted earlier that when a particular value of Y is observed, a reasonable estimate for the true value of X is that value of x which maximizes $p(x|y)$. Since the conditional density function is a maximum (with respect to x) when the exponent is a *minimum*, it follows that this value of x can be determined by equating the derivative of the exponent to zero. Thus,

$$2x - 2(y - b\sigma_N^2) = 0$$

or

$$x = y - b\sigma_N^2 \qquad\qquad (3\text{-}20)$$

is the location of the maximum, *provided that* $y - b\sigma_N^2 > 0$. Otherwise, there is no point of zero slope on $p(x|y)$ and the largest value occurs at $x = 0$. Suppose, therefore, that the value $Y = y_1$ is observed. Then, if $y_1 > b\sigma_N^2$, the appropriate estimate for X is $\hat{X} = y_1 - b\sigma_N^2$. On the other hand, if $y_1 < b\sigma_N^2$, the appropriate estimate for X is $\hat{X} = 0$.

Note that as the noise gets smaller ($\sigma_N{}^2 \rightarrow 0$), the estimate of X approaches the observed value y_1.

Exercise 3-2

A dc signal having a uniform distribution over the range of 0 to 10 V is measured in the presence of an independent noise voltage having a Gaussian distribution with a zero mean and a variance of $2V^2$. What is the best estimate of (a) the signal voltage if the measured value is $0 < y_0 < 10$; (b) the signal voltage if $y_0 < 0$; and (c) the noise voltage if the measured value is $y_0 = 5V$.

Answer: $0, y_0, 0$

3-3   Statistical Independence

The concept of statistical independence was introduced earlier in connection with discrete events, but is equally important in the continuous case. Random variables that arise from different physical sources are almost always statistically independent. For example, the random thermal voltage generated by one resistor in a circuit is in no way related to the thermal voltage generated by another resistor. Statistical independence may also exist when the random variables come from the same source but are defined at greatly different times. For example, the thermal voltage generated in a resistor tomorrow almost certainly does not depend upon the voltage today. When two random variables are statistically independent, a knowledge of one random variable gives no information about the value of the other.

The joint probability density function for statistically independent random variables can always be factored into the two marginal density functions. Thus, the relationship

$$p(x,y) = p_X(x)p_Y(y) \tag{3-21}$$

can be used as a definition for statistical independence, since it can be shown that this factorization is both a necessary and sufficient condition. As an example, this condition is satisfied by the joint density function given in (3-3). Hence, these two random variables are statistically independent.

One of the consequences of statistical independence concerns the correlation defined by (3-5). Because the joint density function is factorable, (3-5) can be written as

$$E[XY] = \int_{-\infty}^{\infty} xp_X(x) \, dx \int_{-\infty}^{\infty} yp_Y(y) \, dy$$
$$= E[X]E[Y] = \bar{X}\bar{Y} \tag{3-22}$$

Hence, the expected value of the product of two statistically independent random variables is simply the product of their mean values.   The result will be zero, of course, if *either* random variable has zero mean.

Another consequence of statistical independence is that conditional probability density functions become marginal density functions.   For example, from (3-12)

$$p(x|y) = \frac{p(x,y)}{p_Y(y)}$$

but if X and Y are statistically independent the joint density function is factorable and this becomes

$$p(x|y) = \frac{p_X(x)p_Y(y)}{p_Y(y)} = p_X(x)$$

Similarly,

$$p(y|x) = \frac{p(x,y)}{p_X(x)} = \frac{p_X(x)p_Y(y)}{p_X(x)} = p_Y(y)$$

3-4   Correlation Between Random Variables

As noted above, one of the important applications of joint probability density functions is that of specifying the *correlation* of two random variables; that is, whether one random variable depends in any way upon another random variable.

If two random variables X and Y have possible values x and y, then the expected value of their product is known as the correlation, defined in (3-5) as

$$E[XY] = \int_{-\infty}^{\infty} \int_{-\infty}^{\infty} xy p(x,y) \, dx \, dy = \bar{X}\bar{Y} \qquad \textbf{(3-5)}$$

If both of these random variables have nonzero means, then it is frequently more convenient to find the correlation with the mean values subtracted out.   Thus,

$$E[(X - \bar{X})(Y - \bar{Y})] = \overline{(X - \bar{X})(Y - \bar{Y})}$$
$$= \int_{-\infty}^{\infty} \int_{-\infty}^{\infty} (x - \bar{X})(y - \bar{Y}) p(x,y) \, dx \, dy \qquad \textbf{(3-23)}$$

This is known as the *covariance*, by analogy to the variance of a single random variable.

If it is desired to express the degree to which two random variables are correlated without regard to the magnitude of either one, then the *correlation coefficient* or *normalized covariance* is the appropriate quantity. The correlation coefficient, which is denoted by ρ, is defined as

$$\rho = E\left[\frac{X - \bar{X}}{\sigma_X}\right]\left[\frac{Y - \bar{Y}}{\sigma_Y}\right] = \int_{-\infty}^{\infty} \int_{-\infty}^{\infty} \frac{x - \bar{X}}{\sigma_X} \cdot \frac{y - \bar{Y}}{\sigma_Y} p(x,y) \, dx \, dy$$
$$\textbf{(3-24)}$$

Note that each random variable has its mean subtracted out and is divided by its standard deviation. The resulting random variable is often called the *standardized variable* and is one with zero mean and unit variance.

In order to investigate some of the properties of ρ, define the standardized variables ξ and η as

$$\xi = \frac{X - \bar{X}}{\sigma_X} \qquad \eta = \frac{Y - \bar{Y}}{\sigma_Y}$$

$$\bar{\xi} = 0 \qquad \bar{\eta} = 0$$

$$\sigma_\xi^2 = 1 \qquad \sigma_\eta^2 = 1$$

Then,

$$\rho = E[\xi\eta]$$

Now look at

$$E[(\xi \pm \eta)^2] = E[\xi^2 \pm 2\xi\eta + \eta^2] = 1 \pm 2\rho + 1$$
$$= 2(1 \pm \rho)$$

Since $(\xi \pm \eta)^2$ is always positive, its expected value must also be positive, so that

$$2(1 \pm \rho) \geq 0$$

Hence, ρ can never have a magnitude greater than one and thus

$$-1 \leq \rho \leq 1$$

If X and Y are statistically independent, then

$$\rho = E[\xi\eta] = \bar{\xi}\bar{\eta} = 0$$

since both ξ and η are zero mean. Thus, the correlation coefficient for statistically independent random variables is always zero. The converse is not necessarily true, however. A correlation coefficient of zero does not automatically mean that X and Y are statistically independent unless they are Gaussian, as will be seen.

Although the correlation coefficient can be defined for any pair of random variables, it is particularly useful for random variables that are individually and jointly Gaussian. In these cases the joint probability density function can be written as

$$p(x,y) = \frac{1}{2\pi\sigma_X\sigma_Y \sqrt{1 - \rho^2}}$$

$$\exp\left\{\frac{-1}{2(1-\rho^2)}\left[\frac{(x-\bar{X})^2}{\sigma_X^2} + \frac{(y-\bar{Y})^2}{\sigma_Y^2} - \frac{2(x-\bar{X})(y-\bar{Y})\rho}{\sigma_X\sigma_Y}\right]\right\} \qquad \text{(3-25)}$$

Note that when $\rho = 0$, this reduces to

$$p(x,y) = \frac{1}{2\pi\sigma_X\sigma_Y} \exp\left\{-\frac{1}{2}\left[\frac{(x-\bar{X})^2}{\sigma_X^2} + \frac{(y-\bar{Y})^2}{\sigma_Y^2}\right]\right\}$$
$$= p_X(x)p_Y(y)$$

which is the form for statistically independent Gaussian random variables. Hence, $\rho = 0$ does imply statistical independence in the Gaussian case.

It is also of interest to use the correlation coefficient to express some results for general random variables. For example, from the definitions of the standardized variables it follows that

$$X = \sigma_X \xi + \bar{X} \qquad \text{and} \qquad Y = \sigma_Y \eta + \bar{Y}$$

and, hence,

$$\bar{XY} = E[(\sigma_X \xi + \bar{X})(\sigma_Y \eta + \bar{Y})] = E[\sigma_X \sigma_Y \xi \eta + \bar{X} \sigma_Y \eta + \bar{Y} \sigma_X \xi + \bar{X}\bar{Y}]$$
$$= \rho \sigma_X \sigma_Y + \bar{X}\bar{Y} \tag{3-26}$$

As a further example, consider

$$E[(X \pm Y)^2] = E[X^2 \pm 2XY + Y^2] = \overline{X^2} \pm 2\bar{X}\bar{Y} + \overline{Y^2}$$
$$= \sigma_X{}^2 + (\bar{X})^2 \pm 2\rho\sigma_X\sigma_Y \pm 2\bar{X}\bar{Y} + \sigma_Y{}^2 + (\bar{Y})^2$$
$$= \sigma_X{}^2 + \sigma_Y{}^2 \pm 2\rho\sigma_X\sigma_Y + (\bar{X} \pm \bar{Y})^2$$

Since the last term is just the square of the mean of $(X \pm Y)$, it follows that the variance of $(X \pm Y)$ is

$$[\sigma_{(X \pm Y)}]^2 = \sigma_X{}^2 + \sigma_Y{}^2 \pm 2\rho\sigma_X\sigma_Y \tag{3-27}$$

Note that when random variables are uncorrelated ($\rho = 0$), the variance of sum or difference is the sum of the variances.

Exercise 3-4

Three random variables X, Y, Z are added together to form a new random variable, $W = X + Y + Z$. The normalized correlation coefficients among the random variables are $\rho_{XY} = 0$, $\rho_{XZ} = \frac{1}{2}$, and $\rho_{YZ} = -\frac{1}{2}$. The means and variances of the random variables are $\bar{X} = 1$, $\bar{Y} = 1$, $\bar{Z} = -1$, and $\sigma_X{}^2 = \sigma_Y{}^2 = \sigma_Z{}^2 = 1$. Find (a) the mean of W and (b) the variance of W.

Answer: 3, 1

3-5 Density Function of the Sum of Two Random Variables

The above example illustrates that the mean and variance associated with the sum (or difference) of two random variables can be determined from a knowledge of the individual means and variances and the correlation coefficient without any regard to the probability density functions of the random variables. A more difficult question, however, pertains to the probability density function of the sum of two random variables. The only situation of this sort that will be considered here is the one in which the two random variables are statistically independent. The more general case is beyond the scope of the present discussion.

Let X and Y be statistically independent random variables with density functions of $p_X(x)$ and $p_Y(y)$, and let the sum be

$$Z = X + Y$$

It is desired to obtain the probability density function of Z, $p_Z(z)$. The situation is best illustrated graphically as shown in Figure 3-4. The

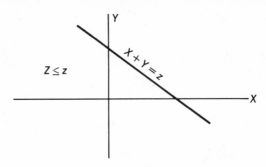

Figure 3-4   Showing the region for $X + Y = Z \leq z$.

probability distribution function for Z is just

$$P_Z(z) = \mathrm{Pr}\,(Z \leq z) = \mathrm{Pr}\,(X + Y \leq z)$$

and can be obtained by integrating the joint density function, $p(x,y)$, over the region *below* the line, $x + y = z$. For every fixed y, x must be such that $-\infty < x < z - y$. Thus,

$$P_Z(z) = \int_{-\infty}^{\infty} \int_{-\infty}^{z-y} p(x,y)\ dx\ dy \qquad \text{(3-28)}$$

For the special case in which X and Y are statistically independent, the joint density function is factorable and (3-28) can be written as

$$P_Z(z) = \int_{-\infty}^{\infty} \int_{-\infty}^{z-y} p_X(x)p_Y(y)\ dy\ dy$$
$$= \int_{-\infty}^{\infty} p_Y(y) \int_{-\infty}^{z-y} p_X(x)\ dx\ dy$$

The probability density function of Z is obtained by differentiating $P_Z(z)$ with respect to z. Hence,

$$p_Z(z) = \frac{dP_Z(z)}{dz} = \int_{-\infty}^{\infty} p_Y(y)p_X(z - y)\ dy \qquad \text{(3-29)}$$

since z appears only in the upper limit of the second integral. Thus, the probability density function of Z is simply the *convolution* of the density functions of X and Y.

It should also be clear that (3-28) could have been written equally well as

$$P_Z(z) = \int_{-\infty}^{\infty} \int_{-\infty}^{z-x} p(x,y)\ dy\ dx$$

and the same procedure would lead to

$$p_Z(z) = \int_{-\infty}^{\infty} p_X(x) p_Y(z - x) \, dx \qquad \textbf{(3-30)}$$

Hence, just as in the case of system analysis, there are two equivalent forms for the convolution integral.

As a simple example of this procedure, consider the two density func-

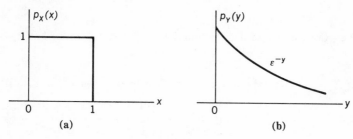

Figure 3-5 Density functions for two random variables.

tions shown in Figure 3-5. These may be expressed analytically as

$$\begin{aligned} p_X(x) &= 1 && 0 \le x \le 1 \\ &= 0 && \text{elsewhere} \end{aligned}$$

and

$$\begin{aligned} p_Y(y) &= \epsilon^{-y} && y \ge 0 \\ &= 0 && y < 0 \end{aligned}$$

The convolution must be carried out in two parts, depending on whether z is greater or less than one. The appropriate diagrams, based on (3-30), are sketched in Figure 3-6. When $0 < z \le 1$, the convolution integral becomes

$$p_Z(z) = \int_0^z (1)\epsilon^{-(z-x)} \, dx = 1 - \epsilon^{-z} \qquad 0 < z \le 1$$

When $z > 1$, the integral is

$$p_Z(z) = \int_0^1 (1)\epsilon^{-(z-x)} \, dx = (\epsilon - 1)\epsilon^{-z} \qquad 1 < z < \infty$$

When $z < 0$, $p_Z(z) = 0$ since both $p_X(x) = 0$, $x < 0$ and $p_Y(y) = 0$, $y < 0$. The resulting density function is sketched in Figure 3-6(c).

It is also of interest to consider the case of the sum of two independent Gaussian random variables. Thus let

$$p_X(x) = \frac{1}{\sqrt{2\pi}\,\sigma_X} \exp\left[\frac{-(x - \bar{X})^2}{2\sigma_X^2}\right]$$

and

$$p_Y(y) = \frac{1}{\sqrt{2\pi}\,\sigma_Y} \exp\left[\frac{-(y - \bar{Y})^2}{2\sigma_Y^2}\right]$$

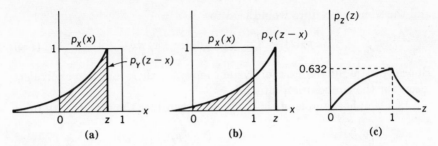

Figure 3-6 Convolution of density functions: (a) $0 < z \leq 1$, (b) $1 < z < \infty$, and (c) $p_Z(z)$.

Then if $Z = X + Y$, the density function for z is [based on (3-30)]

$$p_Z(z) = \frac{1}{2\pi\sigma_X\sigma_Y} \int_{-\infty}^{\infty} \exp\left[\frac{-(x - \bar{X})^2}{2\sigma_X^2}\right] \exp\left[\frac{-(z - x - \bar{Y})^2}{2\sigma_Y^2}\right] dx$$

It is left as an exercise for the student to verify that the result of this integration is

$$p_Z(z) = \frac{1}{\sqrt{2\pi(\sigma_X^2 + \sigma_Y^2)}} \exp\left\{\frac{-[z - (\bar{X} + \bar{Y})]^2}{2(\sigma_X^2 + \sigma_Y^2)}\right\} \tag{3-31}$$

This result clearly indicates that the sum of two *independent* Gaussian random variables is still Gaussian with a mean that is the sum of the means and a variance that is the sum of the variances. It should also be apparent that by adding more random variables the sum is still Gaussian. Thus, the sum of any number of independent Gaussian random variables is still Gaussian. Density functions that exhibit this property are said to be *reproducible;* the Gaussian case is one of a very limited class of density functions that are reproducible. Although it will not be proven here, it can likewise be shown that the sum of *correlated* Gaussian random variables is also Gaussian with a mean that is the sum of the means and a variance that can be obtained from (3-27).

3-6   The Characteristic Function

It was shown in the previous section that the probability density function of the sum of two independent random variables could be obtained by convolving the individual density functions. When more than two random variables are summed, the resulting density function can obviously be obtained by repeating the convolution until every random variable has been taken into account. Since this is a lengthy and tedious procedure, it is natural to inquire if there is some easier way.

When convolution arises in system and circuit analysis, it is well known that transform methods can be used to simplify the computation since convolution then becomes a simple multiplication of the transforms. Repeated convolution is accomplished by multiplying more transforms together. Thus, it seems reasonable to try to use transform methods when dealing with density functions. This section will discuss how to do it.

The *characteristic function* of a random variable X is defined to be

$$\phi(u) = E[\epsilon^{juX}] \tag{3-32}$$

and this expected value can be obtained from

$$\phi(u) = \int_{-\infty}^{\infty} p(x)\epsilon^{jux} \, dx \tag{3-33}$$

The right side of (3-33) is (except for a minus sign in the exponent) the Fourier transform of the density function $p(x)$. The difference in sign for the characteristic function is traditional rather than fundamental, and makes no essential difference in the application or properties of the transform. By analogy to the inverse Fourier transform, the density function can be obtained from

$$p(x) = \frac{1}{2\pi} \int_{-\infty}^{\infty} \phi(u)\epsilon^{-jux} \, du \tag{3-34}$$

In order to illustrate one application of characteristic functions, consider once again the problem of finding the probability density function of the sum of two independent random variables X and Y, where $Z = X + Y$. The characteristic functions for these random variables are

$$\phi_X(u) = \int_{-\infty}^{\infty} p_X(x)\epsilon^{jux} \, dx$$

and

$$\phi_Y(u) = \int_{-\infty}^{\infty} p_Y(y)\epsilon^{jux} \, dy$$

Since convolution corresponds to multiplication of transforms (characteristic functions) it follows that the characteristic function of Z is

$$\phi_Z(u) = \phi_X(u)\phi_Y(u)$$

The resulting density function for Z becomes

$$p_Z(z) = \frac{1}{2\pi} \int_{-\infty}^{\infty} \phi_X(u)\phi_Y(u)\epsilon^{-juz} \, du \tag{3-35}$$

This technique can be illustrated by reworking the example of the previous section, in which X was uniformly distributed and Y exponentially distributed. Since

$$p_X(x) = 1 \qquad 0 \leq x \leq 1$$
$$= 0 \qquad \text{elsewhere}$$

the characteristic function is

$$\phi_X(u) = \int_0^1 (1)\epsilon^{jux}\, dx = \frac{\epsilon^{jux}}{ju}\Big|_0^1$$

$$= \frac{\epsilon^{ju} - 1}{ju}$$

Likewise,

$$p_Y(y) = \epsilon^{-y} \qquad y \geq 0$$
$$= 0 \qquad y < 0$$

so that

$$\phi_Y(u) = \int_0^\infty \epsilon^{-y}\epsilon^{juy}\, dy = \frac{\epsilon^{(-1+ju)y}}{(-1+ju)}\Big|_0^\infty = \frac{1}{1 - ju}$$

Hence, the characteristic function of Z is

$$\phi_Z(u) = \phi_X(u)\phi_Y(u) = \frac{\epsilon^{ju} - 1}{ju(1 - ju)}$$

and the corresponding density function is

$$p_Z(z) = \frac{1}{2\pi}\int_{-\infty}^\infty \frac{\epsilon^{ju} - 1}{ju(1 - ju)}\epsilon^{-juz}\, du$$

$$= \frac{1}{2\pi}\int_{-\infty}^\infty \frac{\epsilon^{ju(1-z)}}{ju(1 - ju)}\, du - \frac{1}{2\pi}\int_{-\infty}^\infty \frac{\epsilon^{-juz}}{ju(1 - ju)}\, du$$

$$= 1 - \epsilon^{-z} \qquad \text{when} \qquad 0 < z < 1$$
$$= (\epsilon - 1)\epsilon^{-z} \qquad \text{when} \qquad 1 < z < \infty$$

The integration can be carried out by standard inverse Fourier transform methods or by the use of tables.

Another application of the characteristic function is to find the moments of a random variable. Note that if $\phi(u)$ is differentiated, the result is

$$\frac{d\phi(u)}{du} = \int_{-\infty}^\infty p(x)(jx)\epsilon^{jux}\, dx$$

For $u = 0$, the derivative becomes

$$\frac{d\phi(u)}{du}\Big|_{u=0} \equiv j\int_{-\infty}^\infty xp(x)\, dx = j\bar{X} \qquad \text{(3-36)}$$

Higher order derivatives introduce higher powers of x into the integrand so that the general nth moment can be expressed as

$$\overline{X^n} = E[X^n] = \frac{1}{j^n}\left[\frac{d^n\phi(u)}{du^n}\right]_{u=0} \qquad \text{(3-37)}$$

If the characteristic function is available, this may be much easier than carrying out the required integrations of the direct approach.

There are some fairly obvious extensions of the above results.   For example, (3-35) can be extended to an arbitrary number of independent random variables.   If $X_1, X_2, \ldots, X_n$ are independent and have characteristic functions of $\phi_1(u), \phi_2(u), \ldots, \phi_n(u)$, and if

$$Y = X_1 + X_2 + \cdots + X_n$$

then Y has a characteristic function of

$$\phi_Y(u) = \phi_1(u)\phi_2(u) \cdots \phi_n(u)$$

and a density function of

$$p_Y(y) = \frac{1}{2\pi} \int_{-\infty}^{\infty} \phi_1(u)\phi_2(u) \cdots \phi_n(u)\epsilon^{-juy} \, du \qquad \text{(3-38)}$$

The characteristic function can also be extended to cases in which random variables are not independent.   For example, if X and Y have a joint density function of $p(x,y)$, then they have a joint characteristic function of

$$\phi_{X,Y}(u,v) = E[\epsilon^{j(uX+vY)}] = \int_{-\infty}^{\infty} \int_{-\infty}^{\infty} p(x,y)\epsilon^{j(ux+vy)} \, dx \, dy \qquad \text{(3-39)}$$

The corresponding inversion relation is

$$p(x,y) = \frac{1}{(2\pi)^2} \int_{-\infty}^{\infty} \int_{-\infty}^{\infty} \phi_{XY}(u,v)\epsilon^{-j(ux+vy)} \, du \, dv \qquad \text{(3-40)}$$

The joint characteristic function can be used to find the correlation between the random variables.   Thus, for example,

$$E[XY] = \bar{X}\bar{Y} = -\left[\frac{\partial^2 \phi_{XY}(u,v)}{\partial u \, \partial v}\right]_{u=v=0} \qquad \text{(3-41)}$$

More generally,

$$E[X^iY^k] = \overline{X^iY^k} = \frac{1}{j^{i+k}}\left[\frac{\partial^{i+k}\phi_{XY}(u,v)}{\partial u^i \, \partial v^k}\right]_{u=v=0} \qquad \text{(3-42)}$$

The results given in (3-37), (3-40), and (3-42) are particularly useful in the case of Gaussian random variables since the necessary integration and differentiations can always be carried out.   One of the valuable properties of Gaussian random variables is that moments and correlations of all orders can be obtained from only a knowledge of the first two moments and the correlation coefficient.

Exercise 3-6

The characteristic function of the Bernoulli distribution is

$$\phi(u) = 1 - p + p\epsilon^{ju}$$

Find (a) the mean, (b) the mean square and (c) the 3rd central moment.

Answer: $p,\; p - 3p^2 + 2p^3,\; p$

■ **PROBLEMS**

3-1 Two random variables, X and Y, have a joint probability density function given by

$$p(x,y) = K \qquad 0 < x < 1 \qquad 0 < y < x$$
$$ = 0 \qquad \text{elsewhere}$$

a. Sketch the probability density function $p(x,y)$.
b. Determine and sketch the distribution function $P(x,y)$.
c. What is the value of K?
d. What is the joint probability of the event that $X < \frac{1}{2}$ and $Y > 0$?
e. What is the marginal density function of X, $p_X(x)$?

3-2 For the two random variables of Problem 3-1

a. Find $E[XY]$.
b. Are the two random variables statistically independent?

3-3 A joint probability density function for X and Y has the form

$$p(x,y) = K\epsilon^{-(3x+4y)}, \qquad x > 0,\, y > 0$$
$$ = 0, \qquad\qquad \text{elsewhere}$$

a. What is the value of K?
b. Find the joint probability distribution function, $P(x,y)$.
c. Determine the probability that $0 < X \le 1$ and $0 < Y \le 2$.
d. Find the marginal density functions for X and Y.

3-4 For the two random variables of Problem 3-1, find the conditional probability density functions $p(x|y)$ and $p(y|x)$.

3-5 A rectangle has dimensions that are random variables. The base, X, is a random variable uniformly distributed from 0 to 5. The height, Y, is a random variable uniformly distributed from 0 to X. Find the expected value of the area of the rectangle.

3-6 A random variable X is considered to be a signal and is uniformly distributed over the range 0 to 10. Another random variable N, independent of X, is considered to be noise and is triangularly distributed over the range -10 to 10 (maximum density at 0). The random variable that can be observed is $Y = X + N$.

a. Sketch and label the conditional probability density function, $p(x|y)$, as a function of x, for $y = -5$, 0, and 5.
b. If an observation yields a value of $y = -5$, what is the best estimate of the true value of X?

3-7   Two random variables have zero means and variances of 25 and 36. Their correlation coefficient is 0.4.

a. What is the variance of their sum?
b. What is the variance of their difference?
c. Repeat parts **a.** and **b.** if the correlation coefficient is -0.4.

3-8   Two statistically independent random variables, X and Y, have zero means and variances of $\sigma_X{}^2 = 36$ and $\sigma_Y{}^2 = 64$.   Two new random variables are defined by

$$U = 4X + 3Y$$
$$V = 3X - 4Y$$

Find the correlation coefficient of U and V.

3-9   A random variable X is linearly related to another random variable Y by the relation

$$Y = aX + b$$

Show that the correlation coefficient of X and Y is

$$\rho_{XY} = \frac{a}{|a|}$$

(that is, when a is positive $\rho_{XY} = +1$ and when a is negative $\rho_{XY} = -1$).

3-10   Let θ and ψ be statistically independent random variables, each being uniformly distributed from 0 to 2π.   Define a new random variable by $\phi = \theta + \psi$.

a. Find the probability density function for ϕ.
b. If $\phi = \theta + \psi,$ $\theta + \psi \leq 2\pi$
                     $= \theta + \psi - 2\pi,$ $2\pi < \theta + \psi \leq 4\pi$

find the probability density function for ϕ.

(Note that in part **a.** the sum of two uniformly distributed random variables is triangularly distributed whereas in part **b.** the sum of two uniformly distributed random variables taken mod 2π is uniformly distributed.)

3-11   **a.** Find the characteristic function of a Gaussian random variable with zero mean and variance σ^2.

b. Using the characteristic function, verify the result given in equation (2-18) for the nth central moment of a Gaussian process.

■ REFERENCES

See References for Chapter 1, particularly Beckmann, Papoulis, and Parzen.

4

Random Processes

4-1   Introduction

It was noted in Chapter 2 that a *random process* is a collection of time functions and an associated probability description. The probability description may consist of the marginal and joint probability density functions of all random variables that are point functions of the process at specified time instants. This type of probability description is the only one that will be considered here.

The entire collection of time functions is an *ensemble* and will be designated as $\{x(t)\}$, where any particular member of the ensemble, $x(t)$, is a *sample function* of the ensemble. In general, only one sample function of a random process can ever be observed; the other sample functions represent all of the other possible realizations that might have occurred but did not. An arbitrary sample function is denoted $X(t)$. The values of $X(t)$ at any time t_1 define a random variable denoted as $X(t_1)$ or simply X_1.

The extension of the concepts of random variables to those of random processes is quite simple as far as the mechanics are concerned; in fact, all of the essential ideas have already been considered. A more difficult step, however, is the conceptual one of relating the mathematical representations for random variables to the physical properties of the process.

Hence, the purpose of this chapter is to help clarify this relationship by means of a number of illustrative examples.

One of the first steps in discussing random processes is that of developing an appropriate terminology that can be used as a "short-cut" in the description of the characteristics of any given process. A convenient way of doing this is to use a set of descriptors, arranged in pairs, and to select one name from each pair to describe the process. Those pairs of descriptors that are appropriate in the present discussion are:

1. Continuous; discrete
2. Deterministic; nondeterministic
3. Stationary; nonstationary
4. Ergodic; nonergodic

4-2   Continuous and Discrete Random Processes

These terms normally apply to the possible values of the random variables. A *continuous random process* is one in which random variables such as $X(t_1)$, $X(t_2)$, and so on, can assume *any* value within a specified range of possible values. This range may be finite, infinite, or semi-infinite. Such things as thermal agitation noise in conductors, shot noise in electron tubes or transistors, and wind velocity are examples of continuous random processes. A sketch of a typical sample function and the corresponding probability density function is shown in Figure 4-1.

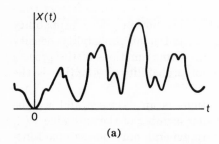
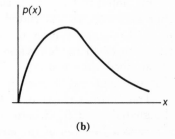

(a)   (b)

Figure 4-1   A continuous random process: (a) typical sample function and (b) probability density function.

In this example, the range of possible values is semi-infinite.

A more precise definition for continuous random processes would be that the probability distribution function is continuous. This would also imply that the density function has no δ functions in it.

A *discrete random process* is one in which the random variables can assume only certain isolated values (possibly infinite in number) and no

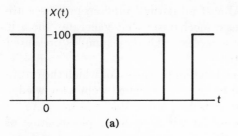
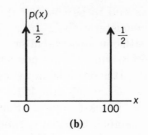

Figure 4-2 A discrete random process: (**a**) typical sample function and
(**b**) probability density function.

other values.   For example, a voltage that is either 0 or 100 because of
random opening and closing of a switch would be a sample function from
a discrete random process.   This is illustrated in Figure 4-2.   Note that
the probability density function contains *only* δ functions.

It is also possible to have *mixed* processes, which have both continuous
and discrete components.   For example, the current flowing in an ideal
rectifier may be zero for one-half the time, as shown in Figure 4-3.   The

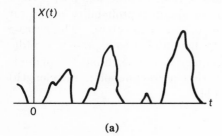
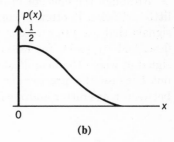

Figure 4-3 A mixed random process: (**a**) typical sample function and
(**b**) probability density function.

corresponding probability density has both a continuous part and a δ
function.

In all of the cases just mentioned, the sample functions are continuous
in *time;* that is, a random variable may be defined for any time.   Situations in which the random variables exist for certain discrete times only
(referred to as *point processes* or *time series*) will not be discussed here.

4-3   Deterministic and Nondeterministic Random Processes

In most of the discussion so far, it has been implied that each sample function is a random function of time and, as such, its future values cannot be
exactly predicted from the observed past values.   Such a random process

is said to be *nondeterministic*.   Almost all natural random processes are nondeterministic, because the basic mechanism that generates them is either unobservable or extremely complex.   All the examples presented in Section 4-2 are nondeterministic.

It is possible, however, to define random processes for which the future values of any sample function can be exactly predicted from a knowledge of the past values.   Such a process is said to be *deterministic*.   As an example, consider a random process for which each sample function of the process is of the form

$$X(t) = A \cos (\omega t + \theta) \tag{4-1}$$

where A and ω are constants and θ is a random variable with a specified probability distribution.   That is, for any one sample function, θ has the same value for all t but different values for the other members of the ensemble.   In this case, the only random variation is over the ensemble— not with respect to time.   It is still possible to define random variables $X(t_1)$, $X(t_2)$, and so on, and to determine probability density functions for them.

Although the concept of deterministic random processes may seem a little artificial, it often is convenient to obtain a probability model for signals that are known except for one or two parameters.   The process described by (4-1), for example, may be suitable to represent a radio signal in which the magnitude and frequency are known but the phase is not because the precise distance (within a fraction of a wavelength) between transmitter and receiver is not.

4-4   Stationary and Nonstationary Random Processes

It has been noted that one can define a probability density function for random variables of the form $X(t_1)$ but so far no mention has been made of the dependence of this density function on the value of time t_1.   If all marginal and joint density functions of the process do not depend upon the choice of time origin, the process is said to be *stationary*.   In this case, all of the mean values and moments discussed previously are constants that do not depend upon the absolute value of time.

If any of the probability density functions do change with the choice of time origin, the process is *nonstationary*.   In this case, one or more of the mean values or moments will also depend on time.   Since the analysis of systems responding to nonstationary random inputs is much more involved than in the stationary case, all future discussion will be limited to the stationary case unless it is specifically stated to the contrary.

In a rigorous sense, there are no stationary random processes that actually exist physically, since any process must have started at some finite time in the past and must presumably stop at some finite time in the future. However, there are many physical situations in which the process does not change appreciably during the time it is being observed. In these cases the stationary assumption leads to a convenient mathematical model, which closely approximates reality.

Determining whether or not the stationary assumption is reasonable for any given situation may not be easy. For nondeterministic processes it depends upon the mechanism of generating the process and upon the time duration over which the process is observed. As a rule of thumb, it is customary to assume stationarity unless there is some obvious change in the source or unless common sense dictates otherwise. For example, the thermal noise generated by the random motion of electrons in a resistor might reasonably be considered stationary under normal conditions. However, if this resistor were being intermittently heated by a current through it, the stationary assumption is obviously false. As another example, it might be reasonable to assume that random wind velocity comes from a stationary source over a period of one hour, say, but common sense indicates that applying this same assumption to a period of one week might be unreasonable.

Deterministic processes are usually stationary only under certain very special conditions. It is customary to assume that these conditions exist, but one must be aware that this is a deliberate choice and not necessarily a natural occurrence. For example, in the case of the random process defined by (4-1), the reader may easily show (by calculating the mean value) that the process may be (and, in fact, is) stationary when θ is uniformly distributed over a range from 0 to 2π, but that it is definitely not stationary when θ is uniformly distributed over a range from 0 to π.

The requirement that all marginal and joint density functions be independent of the choice of time origin is frequently more stringent than is necessary for systems analysis. A more relaxed requirement, which is often adequate, is that the mean value of any random variable, $X(t_1)$, is independent of the choice of t_1 and that the correlation of two random variables, $\overline{X(t_1)X(t_2)}$ depends only upon the time difference, $t_2 - t_1$. Processes that satisfy these two conditions are said to be *stationary in the wide sense*. This wide-sense stationarity is adequate to guarantee that the mean value, mean-square value, variance, and correlation coefficient of any pair of random variables are constants independent of the choice of time origin.

In subsequent discussions of the response of systems to random inputs it will be found that the evaluation of this response is made much easier when the processes may be assumed either strictly stationary or stationary in the wide sense. Since the results are identical for either type of stationarity, it will not be necessary to distinguish between the two in any future discussion.

4-5   Ergodic and Nonergodic Random Processes

Some stationary random processes possess the property that almost every member[1] of the ensemble exhibits the same statistical behavior that the whole ensemble has. Thus, it is possible to determine this statistical behavior by examining only one typical sample function. Such processes are said to be *ergodic*.

For ergodic processes, the mean values and moments can be determined by time averages as well as by ensemble averages. Thus, for example, the nth moment is given by

$$\overline{X^n} = \int_{-\infty}^{\infty} x^n p(x)\, dx = \lim_{T \to \infty} \frac{1}{2T} \int_{-T}^{T} X^n(t)\, dt \qquad \text{(4-2)}$$

It should be emphasized, however, that this condition cannot exist unless the process is stationary. Thus, ergodic processes are also stationary processes.

A process that does not possess the property of (4-2) is *nonergodic*. All nonstationary processes are nonergodic, but it is also possible for stationary processes to be nonergodic. For example, consider sample functions of the form

$$X(t) = Y \cos (\omega t + \theta) \qquad \text{(4-3)}$$

where ω is a constant, Y is a random variable (with respect to the ensemble), and θ is a random variable that is uniformly distributed over 0 to 2π, with θ and Y being statistically independent. This process can be shown to be stationary but nonergodic, since Y is a constant in any one sample function but is different for different sample functions.

It is generally difficult, if not impossible, to prove that ergodicity is a reasonable assumption for any physical process, since only one sample function of the process can be observed. Nevertheless, it is customary to assume ergodicity unless there are compelling physical reasons for not doing so.

4-6   Measurement of Process Parameters

The statistical parameters of a random process are the sets of statistical parameters (such as mean, mean-square, and variance) associated with the $X(t)$ random variables at various times t. In the case of a stationary process these parameters are the same for all such random variables, and, hence, it is customary to consider only one set of parameters.

---

[1] The term "almost every member" implies that a set of sample functions having total probability of zero may not exhibit the same behavior as the rest of the ensemble. But having zero probability does *not* mean that such a sample function is impossible.

A problem of considerable practical importance is that of estimating the process parameters from the observations of a single sample function (since one sample function of finite length is all that is ever available). Because there is only one sample function it is not possible to make an ensemble average in order to obtain estimates of the parameters.   The only alternative, therefore, is to make a time average.   If the process is ergodic, this should be a reasonable approach because a time average (over infinite time) is equivalent to an ensemble average, as indicated by (4-2).

Consider first the problem of estimating the mean value of an ergodic random process $\{x(t)\}$.   This estimate will be designated as $\hat{X}$ and will be computed from a finite time average.   Thus, for an arbitrary member of the ensemble, let

$$\hat{X} = \frac{1}{T} \int_0^T X(t)\, dt \qquad \textbf{(4-4)}$$

It should be noted that although $\hat{X}$ is a single number in any one experiment, it is also a random variable, since a different number would be obtained if a different time interval were used or if a different sample function had been observed.   Thus, $\hat{X}$ will not be identically equal to the true mean value $\bar{X}$, but if the measurement is to be useful it should be close to this value.   Just how close it is likely to be will be discussed below.

Since $\hat{X}$ is a random variable, it has a mean value and a variance. If $\hat{X}$ is to be a good estimate of $\bar{X}$, then the mean value of $\hat{X}$ should be equal to $\bar{X}$ and the variance should be small.   From (4-4) the mean value of $\hat{X}$ is

$$\begin{aligned} E[\hat{X}] &= E\left[\frac{1}{T} \int_0^T X(t)\, dt\right] = \frac{1}{T} \int_0^T E[X(t)]\, dt \\ &= \frac{1}{T} \int_0^T \bar{X}\, dt = \frac{1}{T}\left[\bar{X}t \, \Big|_0^T\right] = \bar{X} \end{aligned} \qquad \textbf{(4-5)}$$

The interchange of expectation and integration is permissible in this case and represents a common type of operation.   The conditions where such interchanges are possible will be discussed in more detail in Chapter 7.   It is clear from (4-5) that $\hat{X}$ has the proper mean value.

The evaluation of the variance of $\hat{X}$ is considerably more involved and will not be considered for the case of continuous time integration, although it will be found for the discrete time case considered next.   It is sufficient to note here that the variance turns out to be proportional to $1/T$.   Thus, a better estimate of the mean is found by averaging the sample function over a longer time interval.   As T approaches infinity, the variance approaches zero and the estimate becomes equal with probability one to the true mean, as it must for an ergodic process.

As a practical matter, the integration required by (4-4) can seldom be carried out analytically because $X(t)$ cannot be expressed in an explicit mathematical form. The alternative is to perform numerical integration upon samples of $X(t)$ observed at equally spaced time instants. Thus, if $X_1 = X(\Delta t)$, $X_2 = X(2\Delta t)$, . . . , $X_N = X(N\Delta t)$, then the estimate of $\bar{X}$ may be expressed as

$$\hat{X} = \frac{1}{N} \sum_{i=1}^{N} X_i \tag{4-6}$$

This is the discrete time counterpart of (4-4).

The estimate $\hat{X}$ is still a random variable and has an expected value of

$$E[\hat{X}] = E\left[\frac{1}{N} \sum_{i=1}^{N} X_i\right] = \frac{1}{N} \sum_{i=1}^{N} E[X_i]$$
$$= \frac{1}{N} \sum_{i=1}^{N} \bar{X} = \bar{X} \tag{4-7}$$

Hence, the estimate still has the proper mean value.

In order to evaluate the variance of $\hat{X}$ it will be assumed that the observed samples are spaced far enough apart in time so that they are statistically independent. This assumption is made for convenience at this point; a more general derivation can be made after considering the material in Chapter 5. The mean-square value of $\hat{X}$ can be expressed as

$$E[(\hat{X})^2] = E\left[\frac{1}{N^2} \sum_{i=1}^{N} \sum_{j=1}^{N} X_i X_j\right] = \frac{1}{N^2} \sum_{i=1}^{N} \sum_{j=1}^{N} E[X_i X_j] \tag{4-8}$$

where the double summation comes from the product of two summations. Since the sample values have been assumed to be statistically independent, it follows that

$$E[X_i X_j] = \overline{X^2} \qquad i = j$$
$$= (\bar{X})^2 \qquad i \neq j$$

Thus,

$$E[(\hat{X})^2] = \frac{1}{N^2} [N\overline{X^2} + (N^2 - N)(\bar{X})^2] \tag{4-9}$$

This results from the fact that the double summation of (4-8) contains N^2 terms all together, but only N of these correspond to $i = j$. Equation (4-9) can be written as

$$E[(\hat{X})^2] = \frac{1}{N} \overline{X^2} + \left(1 - \frac{1}{N}\right)(\bar{X})^2$$
$$= \frac{1}{N} \sigma_X^2 + (\bar{X})^2 \tag{4-10}$$

The variance of $\hat{X}$ can now be written as

$$\text{Var}(\hat{X}) = E[(\hat{X})^2] - \{E[\hat{X}]\}^2 = \frac{1}{N}\sigma_X^2 + (\bar{X})^2 - (\bar{X})^2$$

$$= \frac{1}{N}\sigma_X^2$$

(4-11)

This result says that the variance of the estimate of the mean value is simply $1/N$ times the variance of the process. Thus, the quality of the estimate can be made better by averaging a larger number of samples.

As an illustration of the above result, suppose it is desired to estimate the *variance* of a zero-mean Gaussian random process by passing it through a square law device and estimating the *mean value* of the output. Suppose it is also desired to find the number of sample values that must be averaged in order to be assured that the standard deviation of the resulting estimate is less than 10 percent of the true mean value.

Let the observed sample function of the zero-mean Gaussian process be $Y(t)$ and have a variance of σ_Y^2. After this sample function is squared, it is designated as $X(t)$. Thus,

$$X(t) = Y^2(t)$$

From (2-18) it follows that

$$\bar{X} = E[Y^2] = \sigma_Y^2$$
$$\overline{X^2} = E[Y^4] = 3\sigma_Y^4$$

Hence,

$$\sigma_X^2 = \overline{X^2} - (\bar{X})^2 = 3\sigma_Y^4 - \sigma_Y^4 = 2\sigma_Y^4$$

It is clear from this that an estimate of $\bar{X}$ is also an estimate of σ_Y^2. Furthermore, the variance of the estimate of $\bar{X}$ must be $0.01(\bar{X})^2 = 0.01\sigma_Y^4$ to meet the requirement of an error of less than 10 percent. From (4-11)

$$\text{Var}(\hat{X}) = \frac{1}{N}\sigma_X^2 = \frac{1}{N}(2\sigma_Y^4) = 0.01\sigma_Y^4$$

Thus, $N = 200$ statistically independent samples are required to achieve the desired accuracy.

The preceding not only illustrates the problems in estimating the mean value of a random process, but also indicates how the variance of a zero-mean process might be estimated. The same general procedures can obviously be extended to estimate the variance of a nonzero-mean random process. This is considered in the following exercise.

Exercise 4-6

Let $X(t)$ be any sample function from an ergodic random process having a mean value of $\bar{X}$ and a variance of σ_X^2. Show that σ_X^2 can be estimated by

$$\hat{\sigma}_X{}^2 = \frac{1}{N-1} \sum_{i=1}^{N} X_i{}^2 - \frac{N}{N-1} (\hat{\bar{X}})^2$$

where $(\hat{\bar{X}})$ is the extimate of $\bar{X}$ given by (4-6), and that this estimate has the property

$$E[\hat{\sigma}_X{}^2] = \sigma_X{}^2$$

Explain why the averaging factor is $1/(N-1)$ rather than $1/N$.

▬ PROBLEMS

4-1   In system shown in (a), the switch is initially in position 1.   At $t = 0$, a coin is flipped and the switch is moved to position 2 if a head results and is left on 1 if a tail results.   At $t = 1$, the coin is flipped again and

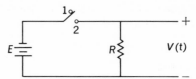

the switch positioned according to the same rule.   This process is repeated at $t = 2$.   At $t = 3$ the switch is assigned to position 1 and remains there.   As an example, if the three coin tosses result in HHT, the corresponding sample function would be as shown in (b).

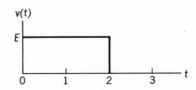

a. Sketch all possible sample functions that belong to the random process described above.
b. What is the probability that any particular sample function will occur?
c. Let a random variable X be defined as $v(.5)$.   What are the possible values that this random variable can assume?   Is this a continuous random variable?

4-2   Classify each of the following random processes according to the four pairs of descriptors listed in Section 4-1.   The basis for the classification should be whether it provides a *useful approximate probability model* for the process and not whether it is rigorously correct.

a. Emission rate (particles per second) of alpha particles observed over a one-hour period from a radioactive source having a half life of 10 years. Observed over a 10-year period.

b. A process for which the random variable is the number of telephone calls going through a telephone exchange at any one time and observed over a 10-minute period. Observed over a 24-hour period. Observed over a 360-day period.

c. A process having sample functions of the form

$$X(t) = at + Y$$

where a is a constant and Y is a random variable.

d. A process having sample functions of the form

$$X(t) = Y \cos \omega_0 t$$

where Y is a random variable and ω_0 is a constant.

e. A process having sample functions of the form $X(t) = Y$, where Y is a random variable.

4-3 Consider a random process having periodic sample functions as shown in which t_0 is a random variable uniformly distributed between 0 and T.   Both A and T are constants.

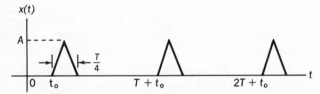

a. Find and sketch the probability distribution function $P(x)$.
b. Find and sketch the probability density function $p(x)$.
c. Find $\bar{X}$, $\overline{X^2}$, and σ^2.
d. Find $\langle x \rangle$ and $\langle x^2 \rangle$, where $\langle \cdot \rangle$ implies a time average.
e. Is this process stationary?   Ergodic?

4-4 Consider a random process having periodic sample functions as shown in which t_0 is a random variable that is uniformly distributed between 0 and T, and Y is a random variable that takes on the values of ± 1 with equal probability and is statistically independent of t_0.   The period T is a constant.

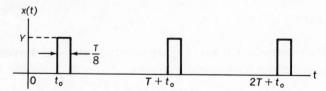

a. Classify this process according to the four pairs of descriptors.

b. Find and sketch the probability density function, $p(x)$, for this process.
c. Calculate $\bar{X}$ and $\overline{X^2}$.
d. Calculate $\langle x \rangle$ and $\langle x^2 \rangle$.
e. Is this process stationary?   Ergodic?

4-5   Ten measurements of a voltage from a Gaussian random process having a mean value of 200 and a variance of 49 are shown below

207	198
202	197
184	213
204	191
206	201

a. Compute the *estimate* of the mean value of the random process.
b. What is the expected value of the estimate of the mean value?
c. What is the variance of the estimate of the mean value?
d. Compute the *estimate* of the variance of the random process.
e. Why are the estimates of the mean value and variance of the random process not equal to the true mean value and variance?

4-6   It is known from the central limit theorem that the average of a large number of independent random variables under very general conditions can be approximated by a Gaussian distribution.   If it is assumed that there is no bias in the measurements of Problem 4-5, then the probability density function for the average is approximately

$$p_{\hat{x}}(\hat{x}) = \frac{\sqrt{n}}{\sqrt{2\pi}\,\sigma} \, e^{-n(\hat{x}-\bar{X})^2/2\sigma^2}$$

where $\bar{X}$ is the true mean and σ^2 the true variance of the measurements. Using this expression and the estimate for σ^2 obtained in Problem 4-5:

a. Determine the limits within which it is 99 percent certain that $\bar{X}$ lies.
b. Determine the value of n that would be required to give 99 percent assurance that the measurements are accurate to one part in the fourth place.

■ REFERENCES

See References for Chapter 1, particularly Beckmann, Davenport and Root, Lanning and Battin, and Papoulis.

5

Correlation Functions

5-1 Introduction

The subject of correlation between two random variables was introduced in Section 3-4. Now that the concept of a random process has also been introduced, it is possible to relate these two subjects to provide a statistical (rather than a probabilistic) description of random processes. Although a probabilistic description is the most complete one, since it incorporates all the knowledge that is available about a random process, there are many engineering situations in which this degree of completeness is neither needed nor possible. If the major interest in a random quantity is in its average power, or the way in which that power is distributed with frequency, then the entire probability model is not needed. If the probability distributions of the random quantities are not known, use of the probability model is not even possible. In either case, a partial statistical description, in terms of certain average values, may provide an acceptable substitute for the probability description.

It was noted in Section 3-4 that the *correlation* between two random variables was the expected value of their product. If the two random variables are defined as samples of a random process at two different time instants, then this expected value depends upon how rapidly the time functions can change. We would expect that the random variables would be highly correlated when the two time instants are very close together, because the time function cannot change rapidly enough to be greatly different.   On the other hand, we would expect to find very little correlation between the values of the random variables when the two time instants are widely separated, because almost any change can take place.

The previously defined correlation was simply a number since the random variables were not necessarily defined as being associated with time functions.   In the following case, however, every pair of random variables can be related by the time separation between them, and the correlation will be a function of this separation.   Thus, it becomes appropriate to define a *correlation function* in which the argument is the time separation of the two random variables.   If the two random variables come from the same random process, this function will be known as the *autocorrelation function*.   If they come from different random processes, it will be called the *cross-correlation function*.   We will consider autocorrelation functions first.

If $X(t)$ is a sample function from a random process, and the random variables are defined to be

$$X_1 = X(t_1)$$
$$X_2 = X(t_2)$$

then the autocorrelation function is defined to be

$$R_X(t_1,t_2) = E[X_1 X_2] = \int_{-\infty}^{\infty} dx_1 \int_{-\infty}^{\infty} x_1 x_2 p(x_1,x_2) \, dx_2 \qquad \textbf{(5-1)}$$

This definition is valid for both stationary and nonstationary random processes.   However, our interest is primarily in stationary processes, for which further simplification of (5-1) is possible.   It will be recalled from the previous chapter that for a wide-sense stationary process all such ensemble averages are independent of the time origin.   Accordingly for a wide-sense stationary process,

$$R_X(t_1,t_2) = R_X(t_1 + T, \, t_2 + T)$$
$$= E[X(t_1 + T)X(t_2 + T)]$$

Since this expression is independent of the choice of time origin, we can set $T = -t_1$ to give

$$R_X(t_1,t_2) = R_X(0, \, t_2 - t_1) = E[X(0)X(t_2 - t_1)]$$

It is seen that this expression depends only on the time difference $t_2 - t_1$. Setting this time difference equal to $\tau = t_2 - t_1$ and suppressing the zero in the argument of $R_X(0, \, t_2 - t_1)$, we can rewrite (5-1) as

$$R_X(\tau) = E[X(t_1)X(t_1 + \tau)] \qquad \textbf{(5-2)}$$

This is the expression for the autocorrelation function of a stationary process and depends only on τ and not on the value of t_1.   Because of this lack of dependence on the particular time t_1 at which the ensemble averages are taken, it is common practice to write (5-2) without the subscript; thus,

$$R_X(\tau) = E[X(t)X(t + \tau)]$$

Whenever correlation functions relate to nonstationary processes, since they are dependent on the particular time at which the ensemble average is taken as well as on the time difference between samples, they must be written as $R_X(t_1, t_2)$ or $R_X(t_1, \tau)$. In all cases in this and subsequent chapters, unless specifically stated otherwise, it is assumed that all correlation functions relate to wide-sense stationary random processes.

It is also possible to define a *time autocorrelation* function for a particular sample function as[1]

$$\mathcal{R}_x(\tau) = \lim_{T \to \infty} \frac{1}{2T} \int_{-T}^{T} x(t)x(t + \tau)\, dt = \langle x(t)x(t + \tau) \rangle \qquad \text{(5-3)}$$

For the special case of an *ergodic process*, $\langle x(t)x(t + \tau) \rangle$ is the same for every $x(t)$ and equal to $R_X(\tau)$. That is,

$$\mathcal{R}_x(\tau) = R_X(\tau) \qquad \text{for an ergodic process} \qquad \text{(5-4)}$$

The assumption of ergodicity, where it is not obviously invalid, often simplifies the computation of correlation functions.

From (5-2) it is seen readily that for $\tau = 0$, since $R_X(0) = E[X(t_1)X(t_1)]$, the autocorrelation function is equal to the mean-square value of the process. For values of τ other than $\tau = 0$, the autocorrelation function $R_X(\tau)$ can be thought of as a measure of the similarity of the waveform $X(t)$ and the waveform $X(t + \tau)$. In order to illustrate this point further, let $X(t)$ be a sample function from a zero-mean stationary random process and form the new function

$$Y(t) = X(t) - \rho X(t + \tau)$$

By determining the value of ρ that minimizes the mean-square value of $Y(t)$ we will have a measure of how much of the waveform $X(t + \tau)$ is contained in the waveform $X(t)$. The determination of ρ is made by computing the variance of $Y(t)$, setting the derivative of the variance with respect to ρ equal to zero, and solving for ρ. The operations are as follows:

$$\begin{aligned}
E\{[Y(t)]^2\} &= E\{[X(t) - \rho X(t + \tau)]^2\} \\
&= E\{X^2(t) - 2\rho X(t)X(t + \tau) + \rho^2 X^2(t + \tau)\} \\
\sigma_Y{}^2 &= \sigma_X{}^2 - 2\rho R_X(\tau) + \rho^2 \sigma_X{}^2 \\
\frac{d\sigma_Y{}^2}{d\rho} &= -2R_X(\tau) + 2\rho\sigma_X{}^2 = 0 \qquad\qquad \text{(5-5)} \\
\rho &= \frac{R_X(\tau)}{\sigma_X{}^2}
\end{aligned}$$

---

[1] The symbol $\langle \ \rangle$ is used to denote time averaging.

It is seen from (5-5) that ρ is directly related to $R_X(\tau)$ and is exactly the *correlation coefficient* defined in Section 3-4. The coefficient ρ can be thought of as the fraction of the waveshape of $X(t)$ remaining after τ seconds have elapsed. It must be remembered that ρ was calculated on a statistical basis; and that it is the average retention of waveshape over the ensemble, and not this property in any particular sample function, that is important. As shown previously, the correlation coefficient ρ can vary from $+1$ to -1. For a value of $\rho = 1$, the waveshapes would be identical—that is, completely correlated. For $\rho = 0$, the waveforms would be completely uncorrelated; that is, no part of the waveform $X(t + \tau)$ would be contained in $X(t)$. For $\rho = -1$, the waveshapes would be identical, except for opposite signs; that is, the waveform $X(t + \tau)$ would be the negative of $X(t)$.

For an ergodic process or for nonrandom signals, the foregoing interpretation can be made in terms of average power instead of variance and in terms of the time correlation function instead of the ensemble correlation function.

Since $R_X(\tau)$ is dependent both on the amount of correlation ρ and the variance of the process, $\sigma_X{}^2$, it is not possible to estimate the significance of some particular value of $R_X(\tau)$ without knowing one or the other of these quantities. For example, if the random process has a zero mean and the autocorrelation function has a positive value, the most that can be said is that the random variables $X(t_1)$ and $X(t_1 + \tau)$ probably have the same sign.[2] If the autocorrelation function has a negative value, it is likely that the random variables have opposite signs. If it is nearly zero, the random variables are about as likely to have opposite signs as they are to have the same sign.

5-2   Example: Autocorrelation Function of a Binary Process

The above ideas may be made somewhat clearer by considering, as a special example, a random process having a very simple autocorrelation function. Figure 5-1 shows a typical sample function from a discrete, stationary, zero-mean random process in which only two values, $\pm A$, are possible. The sample function either can change from one value to the other every t_a seconds or remain the same, with equal probability. The time t_0 is a random variable with respect to the ensemble of possible time functions and is uniformly distributed over an interval of length t_a. This means, as far as the ensemble is concerned, that changes in value can occur at any time with equal probability. It is also assumed that the value of $X(t)$ in any one interval is statistically independent of its value in any other interval.

---

[2] This is strictly true only if $p(x_1)$ is symmetrical about the axis $x_1 = 0$.

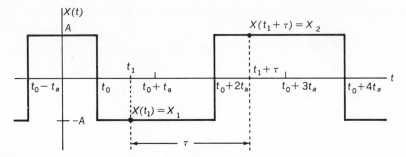

Figure 5-1   A discrete, stationary sample function.

The autocorrelation function of this process will be determined by heuristic arguments rather than by rigorous derivation. In the first place, when $|\tau|$ is larger than t_a, then t_1 and $t_1 + \tau = t_2$ cannot lie in the same interval, and X_1 and X_2 are statistically independent. Since X_1 and X_2 have zero mean, the expected value of their product must be zero, as shown by (3-22); that is,

$$R_X(\tau) = E[X_1 X_2] = \bar{X}_1 \bar{X}_2 = 0 \qquad |\tau| > t_a$$

since $\bar{X}_1 = \bar{X}_2 = 0$. When $|\tau|$ is less than t_a, then t_1 and $t_1 + \tau$ may or may not be in the same interval, depending upon the value of t_0. Since t_0 can be anywhere, with equal probability, then the probability that they do lie in the same interval is proportional to the *difference* between t_a and τ. In particular, for $\tau \geq 0$, it is seen that $t_0 \leq t_1 \leq t_1 + \tau < t_0 + t_a$, which yields $t_1 + \tau - t_a < t_0 \leq t_1$. Hence,

$$\begin{aligned}
\text{Pr } &(t_1 \text{ and } t_1 + \tau \text{ are in the same interval}) \\
&= \text{Pr } [(t_1 + \tau - t_a < t_0 \leq t_1)] \\
&= \frac{1}{t_a}[t_1 - (t_1 + \tau - t_a)] = \frac{t_a - \tau}{t_a}
\end{aligned}$$

since the probability density function for t_0 is just $1/t_a$. When $\tau < 0$, it is seen that $t_0 \leq t_1 + \tau \leq t_1 < t_0 + t_a$, which yields $t_1 - t_a < t_0 \leq t_1 + \tau$. Thus,

$$\begin{aligned}
\text{Pr } &(t_1 \text{ and } t_1 + \tau \text{ are in the same interval}) \\
&= \text{Pr } [((t_1 - t_a) < t_0 \leq (t_1 + \tau))] \\
&= \frac{1}{t_a}[t_1 + \tau - (t_1 - t_a)] = \frac{t_a + \tau}{t_a}
\end{aligned}$$

Hence, in general,

$$\text{Pr } (t_1 \text{ and } t_1 + \tau \text{ are in same interval}) = \frac{t_a - |\tau|}{t_a}$$

When they are in the same interval, the product of X_1 and X_2 is always

A^2; when they are not, the expected product is zero. Hence,

$$R_X(\tau) = A^2 \left[\frac{t_a - |\tau|}{t_a} \right] = A^2 \left[1 - \frac{|\tau|}{t_a} \right] \qquad 0 \leq |\tau| \leq t_a \tag{5-6}$$
$$= 0 \qquad\qquad\qquad\qquad |\tau| \geq t_a$$

This function is sketched in Figure 5-2.

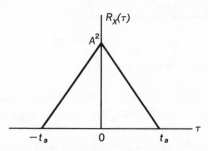

Figure 5-2 Autocorrelation function of the process in Figure 5-1.

It is interesting to consider the physical interpretation of this auto-correlation function in the light of the previous discussion. Note that when $|\tau|$ is small (less than t_a), there is an increased probability that $X(t_1)$ and $X(t_1 + \tau)$ will have the same value, and the autocorrelation function is positive. When $|\tau|$ is greater than t_a, it is equally probable that $X(t_1)$ and $X(t_1 + \tau)$ will have the same value as that they will have opposite values, and the autocorrelation function is zero. For $\tau = 0$, the autocorrelation function yields the mean-square value of A^2.

5-3 Properties of Autocorrelation Functions

There are a few important general properties that are possessed by all autocorrelation functions of stationary random processes. These may be summarized as follows:

1. $R_X(0) = \overline{X^2}$. Hence, the mean-square value of the random process can always be obtained simply by setting $\tau = 0$.

2. $R_X(\tau) = R_X(-\tau)$. The autocorrelation function is an *even* function of τ. It is even because the same set of product values is averaged, regardless of the direction of translation.

3. $|R_X(\tau)| \leq R_X(0)$. The largest value of the autocorrelation function always occurs at $\tau = 0$. There may be other values of τ for which it is just as big (for example, see the periodic case below), but it cannot be larger. This is shown easily by considering

$$E[(X_1 \pm X_2)^2] = E[X_1^2 + X_2^2 \pm 2X_1X_2] \geq 0$$
$$E[X_1^2 + X_2^2] = 2R_X(0) \geq |E(2X_1X_2)| = |2R_X(\tau)|$$

and thus,

$$R_X(0) \geq |R_X(\tau)| \qquad \text{(5-7)}$$

4.   If $X(t)$ has a dc component or mean value, then $R_X(\tau)$ will have a constant component.   For example, if $X(t) = A$, then

$$R_X(\tau) = E[X(t_1)X(t_1 + \tau)] = E[AA] = A^2 \qquad \text{(5-8)}$$

More generally, if $X(t)$ has a mean value *and* a zero mean component $N(t)$ so that

$$X(t) = \bar{X} + N(t)$$

then

$$
\begin{aligned}
R_X(\tau) &= E\{[\bar{X} + N(t_1)][\bar{X} + N(t_1 + \tau)]\} \\
&= E[(\bar{X})^2 + \bar{X}N(t_1) + \bar{X}N(t_1 + \tau) + N(t_1)N(t_1 + \tau)] \quad \text{(5-9)} \\
&= (\bar{X})^2 + R_N(\tau)
\end{aligned}
$$

since

$$E[N(t_1)] = E[N(t_1 + \tau)] = 0$$

Thus, even in this case, $R_X(\tau)$ contains a constant component.

5.   If $X(t)$ has a periodic component, then $R_X(t)$ will also have a periodic component, with the same period.   For example, let

$$X(t) = A \cos (\omega t + \theta)$$

where A and ω are constants and θ is a random variable uniformly distributed over a range of 2π.   That is,

$$
\begin{aligned}
p(\theta) &= \frac{1}{2\pi} \qquad 0 \leq \theta \leq 2\pi \\
&= 0 \qquad \text{elsewhere}
\end{aligned}
$$

Then

$$
\begin{aligned}
R_X(\tau) &= E[A \cos (\omega t_1 + \theta)A \cos (\omega t_1 + \omega \tau + \theta)] \\
&= E\left[\frac{A^2}{2} \cos (2\omega t_1 + \omega \tau + 2\theta) + \frac{A^2}{2} \cos \omega \tau\right] \\
&= \frac{A^2}{2} \int_0^{2\pi} \frac{1}{2\pi} [\cos (2\omega t_1 + \omega \tau + 2\theta) + \cos \omega \tau] \, d\theta \\
&= \frac{A^2}{2} \cos \omega \tau
\end{aligned} \qquad \text{(5-10)}
$$

In the more general case, in which

$$X(t) = A \cos (\omega t + \theta) + N(t)$$

where θ and $N(t_1)$ are statistically independent for all t_1, by the method used in obtaining (5-9), it is easy to show that

$$R_X(\tau) = \frac{A^2}{2} \cos \omega \tau + R_N(\tau) \qquad \text{(5-11)}$$

Hence, the autocorrelation function still contains a periodic component.

6. If $\{X(t)\}$ is ergodic and zero mean, and has no periodic components, then

$$\lim_{|\tau| \to \infty} R_X(\tau) = 0 \qquad \text{(5-12)}$$

For large values of τ, since the effect of past values tends to die out as time progresses, the random variables tend to become statistically independent.

7. Autocorrelation functions cannot have an arbitrary shape. One way of specifying shapes that are permissible is in terms of the Fourier transform of the autocorrelation function. That is, if

$$\mathfrak{F}[R_X(\tau)] = \int_{-\infty}^{\infty} R_X(\tau)\, \epsilon^{-j\omega\tau}\, d\tau$$

then the restriction is

$$\mathfrak{F}[R_X(\tau)] \geq 0 \qquad \text{all } \omega \qquad \text{(5-13)}$$

The reason for this restriction will become apparent after the discussion of spectral density in Chapter 6. Among other things, this restriction precludes the existence of autocorrelation functions with flat tops, vertical sides, or any discontinuity in amplitude.

There is one further point that should be emphasized in connection with autocorrelation functions. Although a knowledge of the joint probability density functions of the random process is sufficient to obtain a unique autocorrelation function, the converse is not true. There may be many different random processes that can yield the same autocorrelation function. Furthermore, as will be shown later, the effect of linear systems on the autocorrelation function of the input can be computed *without* knowing anything about the probability density functions. Hence, the specification of the correlation function of a random process is not equivalent to the specification of the probability density functions and, in fact, represents a considerably smaller amount of information.

Exercise 5-3

An autocorrelation function has the form

$$R_X(\tau) = 100\epsilon^{-10|\tau|} + 100 \cos 10\tau + 100$$

Find the mean, mean-square value, and variance of this process.

Answer: 200, ± 10, 300

5-4   Measurement of Autocorrelation Functions

Since the autocorrelation function plays an important role in the analysis of linear systems with random inputs, an important practical problem is that of determining these functions for experimentally observed random processes. In general, they cannot be calculated from the joint density functions, since these density functions are seldom known. Nor can an ensemble average be made, because there is usually only one sample function from the ensemble available. Under these circumstances, the only available procedure is to calculate a time autocorrelation function for a finite time interval, under the assumption that the process is ergodic.

In order to illustrate this, assume that a particular voltage or current waveform $x(t)$ has been observed over a time interval from 0 to T seconds. It is then possible to define an *estimated* correlation function as for this particular waveform as

$$\hat{R}_x(\tau) = \frac{1}{T - \tau} \int_0^{T-\tau} x(t)x(t + \tau)\, dt \qquad 0 \le \tau \ll T \tag{5-14}$$

Over the ensemble of sample functions, this estimate is a random variable denoted by $\hat{R}_X(\tau)$. Note that the averaging time is $T-\tau$ rather than T because this is the only portion of the observed data in which both $x(t)$ and $x(t + \tau)$ are available.

In most practical cases it is not possible to carry out the integration called for in (5-14) because a mathematical expression for $x(t)$ is not available. An alternative procedure is to approximate the integral by sampling the continuous time function at discrete instants of time and performing the discrete equivalent to (5-14). Thus, if the samples of a particular sample function are taken at time instants of $0, \Delta t, 2\Delta t, \ldots,$ $N\Delta t$, and if the corresponding values of $x(t)$ are $x_0, x_1, x_2, \ldots, x_N$, the discrete equivalent to (5-14) is

$$\hat{R}_x(n\,\Delta t) = \frac{1}{N - n + 1} \sum_{k=0}^{N-n} x_k x_{k+n} \qquad n = 0, 1, 2, \ldots, M \tag{5-15}$$

$$M \ll N$$

This estimate is also a random variable over the ensemble and, as such, is denoted by $\hat{R}_X(n\Delta t)$. Since N is quite large (on the order of several thousand) this operation is best performed by a digital computer, and standard programs for doing this are available.

In order to evaluate the quality of this estimate it is necessary to determine the mean and the variance of $\hat{R}_X(n\Delta t)$, since it is a random variable whose precise value depends upon the particular sample function being used and the particular set of samples taken. The mean is easy enough to obtain since

$$E[\hat{R}_X(n\Delta t)] = E\left[\frac{1}{N-n+1}\sum_{k=0}^{N-n}X_kX_{k+n}\right]$$

$$= \frac{1}{N-n+1}\sum_{k=0}^{N-n}E[X_kX_{k+n}] = \frac{1}{N-n+1}\sum_{k=0}^{N-n}R_X(n\Delta t)$$

$$= R_X(n\Delta t)$$

Thus, the expected value of the estimate is the true value of the auto-correlation function. Such an estimate is said to be *unbiased*.

It is much more difficult to determine the variance of the estimate, and the details of this are beyond the scope of the present discussion. It is possible to show, however, that the variance of the estimate must be smaller than

$$\text{Var }[\hat{R}_X(n\Delta t)] \leq \frac{2}{N}\sum_{k=-M}^{M}R_X^2(k\Delta t) \tag{5-16}$$

As an illustration of what this result means in terms of the number of samples required for a given degree of accuracy, suppose that it is desired to estimate a correlation function of the form shown in Figure 5-2 with 4 points on either side of center ($M = 4$). If an rms error of 5 percent[3] or less is required, then (5-16) implies that (since $t_a = 4\Delta t$)

$$(.05A^2)^2 \leq \frac{2}{N}\sum_{k=-4}^{4}A^4\left[1 - \frac{|k|\,\Delta t}{4\Delta t}\right]^2$$

This can be solved for N to obtain

$$N \geq 2200$$

It is clear that long samples of data and extensive calculations are necessary if accurate estimates of correlation functions are to be made.

Exercise 5-4

For a triangular autocorrelation function of the type shown in Figure 5-2, find the value of N required if it is desired to achieve an rms error of 1 percent or less and if the correlation function is estimated at 5 points on either side of center ($M = 5$).

Answer: (Select one) 6800, 34,000, 68,000

---

[3] This implies that the standard deviation of the estimate should be no greater than 5 percent of the true mean value of the random variable $\hat{R}_X(n\Delta t)$.

5-5   Examples of Autocorrelation Functions

Before going on to consider cross-correlation functions it is worthwhile to look at some typical autocorrelation functions, suggest the circumstances under which they might arise, and list possible applications. This discussion is not intended to be exhaustive but primarily to introduce some ideas.

The triangular correlation function shown in Figure 5-2 is typical of random binary signals in which the switching must occur at uniformly spaced time intervals.   Such a signal arises in many types of communication and control systems in which the continuous signals are sampled at periodic instants of time and the resulting sample amplitudes converted to binary numbers.   The correlation function shown in Figure 5-2 assumed that the random process had a mean value of zero, but this is not always the case.   If, for example, the random signal could assume values of A and 0 (rather than $-A$) then the process has a mean value of $A/2$ and a mean-square value of $A^2/2$.   The resulting autocorrelation function, shown in Figure 5-3, follows from an application of (5-9).

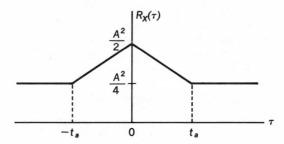

Figure 5-3 Autocorrelation function of a binary process with a mean value.

Not all binary time functions have triangular autocorrelation functions, however.   For example, another common type of binary signal is one in which the switching occurs at randomly spaced instants of time. If all times are equally probable, then the probability density function associated with the duration of each interval is exponential, as shown in Section 2-7.   The resulting autocorrelation function is also exponential, as shown in Figure 5-4.   The usual mathematical representation of such an autocorrelation function is

$$R_X(\tau) = A^2\epsilon^{-\alpha|\tau|} \tag{5-17}$$

where α is the average number of intervals per second.

Binary signals and correlation functions of the type shown in Figure 5-4 frequently arise in connection with radioactive monitoring devices. The randomly occurring pulses at the output of a particle detector are used to trigger a flip-flop circuit that generates the binary signal.   This

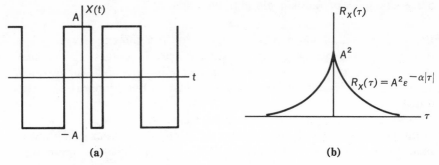

(a)

(b)

Figure 5-4   (a) A binary signal with randomly spaced switching times and (b) the corresponding autocorrelation function.

type of signal is a convenient one for measuring either the average time interval between particles or the average rate of occurrence.

Nonbinary signals can also have exponential correlation functions. For example, if very wideband noise (having almost any probability density function) is passed through a low-pass RC filter, the signal appearing at the output of the filter will have a nearly exponential autocorrelation function.   This result will be shown in detail in Chapter 7.

All of the correlation functions discussed so far have been positive for all values of τ.   This is not necessary, however, and two common types of autocorrelation functions that have negative regions are given by

$$R_X(\tau) = A^2 \epsilon^{-\alpha|\tau|} \cos \beta\tau \qquad (5\text{-}18)$$

and

$$R_X(\tau) = \frac{A^2 \sin \pi\gamma\tau}{\pi\gamma\tau} \qquad (5\text{-}19)$$

and are illustrated in Figure 5-5.   The autocorrelation function of (5-18)

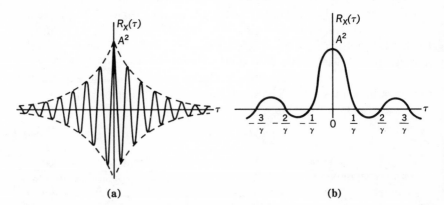

(a)

(b)

Figure 5-5   The autocorrelation functions arising at the outputs of (a) a band-pass filter and (b) an ideal low-pass filter.

arises at the output of the narrow-band bandpass filter whose input is very wideband noise, while that of (5-19) is typical of the autocorrelation at the output of an ideal low-pass filter. Both of these results will be derived in Chapters 6 and 7.

Although there are many other types of autocorrelation functions that arise in connection with signal and system analysis, the few discussed here are the ones most commonly encountered. The student should refer to the properties of autocorrelation functions discussed in Section 5-3 and verify that all these correlation functions possess those properties.

Exercise 5-5

Suppose the binary waveform shown in Figure 5-4(a) has values of A and 0 (instead of $-A$). Sketch and label the resulting autocorrelation function.

5-6   Cross-correlation Functions

It is also possible to consider the correlation between two random variables from different random processes. This situation arises when there is more than one random signal being applied to a system or when one wishes to compare random voltages or currents occurring at different points in the system. If the random processes are jointly stationary in the wide sense, and if sample functions from these processes are designated as $X(t)$ and $Y(t)$, then for two random variables

$$X_1 = X(t_1)$$
$$Y_2 = Y(t_1 + \tau)$$

it is possible to define the *cross-correlation function*

$$R_{XY}(\tau) = E[X_1 Y_2] = \int_{-\infty}^{\infty} dx_1 \int_{-\infty}^{\infty} x_1 y_2 p(x_1, y_2)\, dy_2 \qquad \textbf{(5-20)}$$

The order of subscripts is significant; the second subscript refers to the random variable taken at $(t_1 + \tau)$.[4]

There is also another cross-correlation function that can be defined for the same two instants. Thus, let

$$Y_1 = Y(t_1)$$
$$X_2 = X(t_1 + \tau)$$

and define

$$R_{YX}(\tau) = E[Y_1 X_2] = \int_{-\infty}^{\infty} dy_1 \int_{-\infty}^{\infty} y_1 x_2 p(y_1, x_2)\, dx_2 \qquad \textbf{(5-21)}$$

Note that because both random processes were assumed to be *jointly* sta-

---

[4] This is an arbitrary convention, which is by no means universal with all authors. The definitions should be checked in every case.

tionary, these cross-correlation functions depend only upon the time difference τ.

The *time cross-correlation functions* may be defined as before for a particular pair of sample functions as

$$\mathfrak{R}_{xy}(\tau) = \lim_{T \to \infty} \frac{1}{2T} \int_{-T}^{T} x(t)y(t + \tau)\, dt \qquad \text{(5-22)}$$

and

$$\mathfrak{R}_{yx}(\tau) = \lim_{T \to \infty} \frac{1}{2T} \int_{-T}^{T} y(t)x(t + \tau)\, dt \qquad \text{(5-23)}$$

If the random processes are jointly ergodic, then (5-22) and (5-23) yield the same value for every pair of sample functions. Hence, for ergodic processes,

$$\mathfrak{R}_{xy}(\tau) = R_{XY}(\tau) \qquad \text{(5-24)}$$
$$\mathfrak{R}_{yx}(\tau) = R_{YX}(\tau) \qquad \text{(5-25)}$$

In general, the physical interpretation of cross-correlation functions is no more concrete than that of autocorrelation functions. It is simply a measure of how much these two random variables depend upon one another. In the later study of system analysis, however, the specific cross-correlation function between system input and output will take on a very definite and important physical significance.

5-7   Properties of Cross-correlation Functions

The general properties of all cross-correlation functions are quite different from those of autocorrelation functions. They may be summarized as follows:

1. The quantities $R_{XY}(0)$ and $R_{YX}(0)$ have *no* particular physical significance and do *not* represent mean-square values. It is true, however, that $R_{XY}(0) = R_{YX}(0)$.

2. Cross-correlation functions are not generally even functions of τ. There is a type of symmetry, however, as indicated by the relations

$$R_{YX}(\tau) = R_{XY}(-\tau) \qquad \text{(5-26)}$$

These results follow from the fact that a shift of $Y(t)$ in one direction (in time) is equivalent to a shift of $X(t)$ in the other direction.

3. The cross-correlation function does not necessarily have its maximum value at $\tau = 0$. It can be shown, however, that

$$|R_{XY}(\tau)| \leq [R_X(0)R_Y(0)]^{1/2} \qquad \text{(5-27)}$$

with a similar relationship for $R_{YX}(\tau)$. The maximum of the cross-correlation function can occur anywhere, but it cannot exceed the above value.   Furthermore, it may not achieve this value anywhere.

4.   If the two random processes are statistically independent, then

$$R_{XY}(\tau) = E[X_1, Y_2] = E[X_1]E[Y_2] = \bar{X}\bar{Y}$$
$$= R_{YX}(\tau)$$

(5-28)

If, in addition, *either* process has zero mean, then the cross-correlation function vanishes for all τ.   The converse of this is not necessarily true, however.   The fact that the cross-correlation function is zero and that one process has zero mean does *not* imply that the random processes are statistically independent, except for jointly Gaussian random variables.

5.   If $X(t)$ is a stationary random process and $\dot{X}(t)$ is its derivative with respect to time, the cross-correlation function of $X(t)$ and $\dot{X}(t)$ is given by

$$R_{X\dot{X}}(\tau) = \frac{dR_X(\tau)}{d\tau}$$

(5-29)

in which the right side of (5-29) is the derivative of the autocorrelation function with respect to τ.   This is easily shown by employing the fundamental definition of a derivative

$$\dot{X}(t) = \lim_{\epsilon \to 0} \frac{X(t + \epsilon) - X(t)}{\epsilon}$$

Hence,

$$R_{X\dot{X}}(\tau) = E[X(t)\dot{X}(t + \tau)]$$
$$= E\left\{ \lim_{\epsilon \to 0} \frac{X(t)X(t + \tau + \epsilon) - X(t)X(t + \tau)}{\epsilon} \right\}$$
$$= \lim_{\epsilon \to 0} \frac{R_X(\tau + \epsilon) - R_X(\tau)}{\epsilon} = \frac{dR_X(\tau)}{d(\tau)}$$

The interchange of the limit operation and the expectation is permissible whenever $\dot{X}(t)$ exists.   If the above process is repeated, it is also possible to show that the autocorrelation function of $\dot{X}(t)$ is

$$R_{\dot{X}}(\tau) = R_{\dot{X}\dot{X}}(\tau) = -\frac{d^2 R_X(\tau)}{d\tau^2}$$

(5-30)

where the right side is the second derivative of the basic autocorrelation function with respect to τ.

Exercise 5-7

Prove the result given in (5-27) by considering the value of

$$E\left\{ \left[\frac{X_1}{\sqrt{R_X(0)}} \pm \frac{Y_2}{\sqrt{R_Y(0)}} \right]^2 \right\}$$

5-8   Examples and Applications of Cross-correlation Functions

It was noted previously that one of the applications of cross-correlation functions was in connection with systems with two or more random inputs.   In order to explore this in more detail, consider a random process whose sample functions are of the form

$$Z(t) = X(t) \pm Y(t)$$

in which $X(t)$ and $Y(t)$ are also sample functions of random processes. Then defining the random variables as

$$Z_1 = X_1 \pm Y_1 = X(t_1) \pm Y(t_1)$$
$$Z_2 = X_2 \pm Y_2 = X(t_1 + \tau) \pm Y(t_1 + \tau)$$

the autocorrelation function of $Z(t)$ is

$$
\begin{aligned}
R_Z(\tau) &= E[Z_1 Z_2] = E[(X_1 \pm Y_1)(X_2 \pm Y_2)] \\
&= E[X_1 X_2 + Y_1 Y_2 \pm X_1 Y_2 \pm Y_1 X_2] \\
&= R_X(\tau) + R_Y(\tau) \pm R_{XY}(\tau) \pm R_{YX}(\tau)
\end{aligned}
\tag{5-31}
$$

This result is easily extended to the sum of any number of random variables.   In general, the autocorrelation function of such a sum will be the sum of *all* the autocorrelation functions plus the sum of *all* the cross-correlation functions.

If the two random processes being considered are statistically independent and one of them has zero mean, then both of the cross-correlation functions in (5-31) vanish and the autocorrelation function of the sum is just the sum of the autocorrelation functions.   An example of the importance of this result arises in connection with the extraction of periodic signals from random noise.   Let $X(t)$ be a desired signal sample function of the form

$$X(t) = A \cos (\omega t + \theta) \tag{5-32}$$

where θ is a random variable.   It was shown previously that the autocorrelation function of this process is

$$R_X(\tau) = \frac{1}{2} A^2 \cos \omega \tau \tag{5-33}$$

Next, let $Y(t)$ be a sample function of zero-mean random noise that is statistically independent of the signal and specify that it has an autocorrelation function of the form

$$R_Y(\tau) = B^2 \epsilon^{-\alpha|\tau|} \tag{5-34}$$

The observed quantity is $Z(t)$, which from (5-31) has an autocorrelation function of

$$
\begin{aligned}
R_Z(\tau) &= R_X(\tau) + R_Y(\tau) \\
&= \frac{1}{2} A^2 \cos \omega \tau + B^2 \epsilon^{-\alpha|\tau|}
\end{aligned}
\tag{5-35}
$$

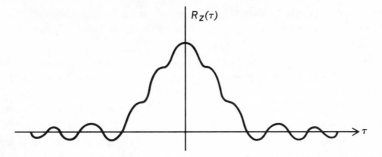

Figure 5-6   Autocorrelation function of sinusoidal signal plus noise.

This function is sketched in Figure 5-6 for a case in which the average noise power, Y^2, is much larger than the average signal power, $\frac{1}{2}A^2$.   It is clear from the sketch that for large values of τ, the autocorrelation function depends mostly upon the signal, since the noise autocorrelation function tends to zero as τ tends to infinity.   Thus, it should be possible to extract tiny amounts of sinusoidal signal from large amounts of noise by using an appropriate method for measuring the autocorrelation function of the received signal plus noise.

Another method of extracting a small known signal from a combination of signal and noise is to perform a cross-correlation operation.   A typical example of this might be a radar system that is transmitting a signal $X(t)$.   The signal that is returned from any target is a very much smaller version of $X(t)$ and has been delayed in time by the propagation time to the target and back.   Since noise is always present at the input to the radar receiver, the total received signal $Y(t)$ may be represented as

$$Y(t) = aX(t - \tau_1) + N(t) \tag{5-36}$$

where a is a number very much smaller than 1, τ_1 is the round-trip delay time of the signal, and $N(t)$ is the receiver noise.   In a typical situation the average power of the returned signal, $aX(t - \tau_1)$, is very much smaller than the average power of the noise, $N(t)$.

The cross-correlation function of the transmitted signal and the total receiver input is

$$\begin{aligned} R_{XY}(\tau) &= E[X(t)Y(t + \tau)] \\ &= E[aX(t)X(t + \tau - \tau_1) + X(t)N(t + \tau)] \\ &= aR_X(\tau - \tau_1) + R_{XN}(\tau) \end{aligned} \tag{5-37}$$

Since the signal and noise are statistically independent and have zero mean (because they are RF bandpass signals), the cross-correlation function between $X(t)$ and $N(t)$ is zero for all values of τ.   Thus, (5-37) becomes

$$R_{XY}(\tau) = aR_X(\tau - \tau_1) \tag{5-38}$$

Remembering that autocorrelation functions have their maximum values at the origin, it is clear that if τ is adjusted so that the measured value of $R_{XY}(\tau)$ is a maximum, then $\tau = \tau_1$ and this value indicates the distance to the target.

The actual measurement of cross-correlation functions can be carried out in much the same way as that suggested for measuring autocorrelation functions in Section 5-4.   This type of measurement is still unbiased when cross-correlation functions are being considered, but the result given in (5-16) for the variance of the estimate is no longer strictly true—particularly if one of the signals contains additive uncorrelated noise, as in the radar example just discussed.   Generally speaking, the number of samples required to obtain a given variance in the estimate of a cross-correlation function is much greater than that required for an autocorrelation function.

5-9   Correlation Matrices for Sampled Functions

A situation in which vector notation is useful in representing a signal arises in the case of a single time function that is sampled at periodic time instants.   If only a finite number of such samples are to be considered, say N, then each sample value can become a component of an $(N \times 1)$ vector.   Thus, if the sampling times are $t_1, t_2, \ldots, t_N$, the vector representing the time function $X(t)$ may be expressed as

$$\mathbf{X} = \begin{bmatrix} X(t_1) \\ X(t_2) \\ \cdot \\ \cdot \\ \cdot \\ X(t_N) \end{bmatrix}$$

If $X(t)$ is a sample function from a random process, then each of the components of the vector $\mathbf{X}$ is a random variable.

It is now possible to define a correlation matrix that is $(N \times N)$ and gives the correlation between every pair of random variables.   Thus,

$$\mathbf{R_x} = E[\mathbf{X}\mathbf{X}^T] = E\begin{bmatrix} X(t_1)X(t_1) & X(t_1)X(t_2) & \cdots & X(t_1)X(t_N) \\ X(t_2)X(t_1) & X(t_2)X(t_2) & & \\ \cdot & & & \\ \cdot & & & \\ \cdot & & & \\ X(t_N)X(t_1) & & \cdots & X(t_N)X(t_N) \end{bmatrix}$$

where $\mathbf{X}^T$ is the transpose of $\mathbf{X}$.   When the expected value of each element of the matrix is taken, that element becomes a particular value of the autocorrelation function of the random process from which $X(t)$ came.

Thus,

$$\mathbf{R_x} = \begin{bmatrix} R_X(t_1,t_1) & R_X(t_1,t_2) & \cdots & R_X(t_1,t_N) \\ R_X(t_2,t_1) & R_X(t_2,t_2) & & \\ & \cdot & & \\ & \cdot & & \\ & \cdot & & \\ R_X(t_N,t_1) & & \cdots & R_X(t_N,t_N) \end{bmatrix} \qquad (5\text{-}39)$$

When the random process from which $X(t)$ came is wide-sense stationary, then all the components of $\mathbf{R_x}$ become functions of time difference only.   If the interval between sample values is Δt, then

$$t_2 = t_1 + \Delta t$$
$$t_3 = t_1 + 2\Delta t$$
$$\cdot$$
$$\cdot$$
$$\cdot$$
$$t_N = t_1 + (N-1)\,\Delta t$$

and

$$\mathbf{R_x} = \begin{bmatrix} R_X[0] & R_X[\Delta t] & \cdots & R_X[(N-1)\,\Delta t] \\ R_X[\Delta t] & R_X[0] & & \\ & \cdot & & \\ & \cdot & & \\ & \cdot & & \\ R_X[(N-1)\,\Delta t] & & \cdots & R_X[0] \end{bmatrix} \qquad (5\text{-}40)$$

where use has been made of the symmetry of the autocorrelation function; that is, $R_X[i\,\Delta t] = R_X[-i\,\Delta t]$.   Note that as a consequence of the symmetry, $\mathbf{R_x}$ is a symmetric matrix (even in the nonstationary case), and that as a consequence of stationarity, the major diagonal (and all diagonals parallel to it) have identical elements.

Although the $\mathbf{R_x}$ just defined is a logical consequence of previous definitions, it is not the most customary way of designating the correlation matrix of a random vector consisting of sample values.   A more common procedure is to define a *covariance matrix*, which contains the variances and covariances of the random variables.   The general covariance between two random variables is defined as

$$E\{[X(t_i) - \bar{X}(t_i)][X(t_j) - \bar{X}(t_j)]\} = \sigma_i \sigma_j \rho_{ij} \qquad (5\text{-}41)$$

where
$\bar{X}(t_i)$ = mean value of $X(t_i)$
$\bar{X}(t_j)$ = mean value of $X(t_j)$
σ_i^2 = variance of $X(t_i)$
σ_j^2 = variance of $X(t_j)$
ρ_{ij} = normalized covariance coefficient of $X(t_i)$ and $X(t_j)$
  = 1, when $i = j$

The covariance matrix is defined as

$$\Lambda_X = E[X - \bar{X})(X^T - \bar{X}^T)] \tag{5-42}$$

where $\bar{X}$ is the mean value of X.   Using the covariance definitions leads immediately to

$$\Lambda_X = \begin{bmatrix} \sigma_1{}^2 & \sigma_1\sigma_2\rho_{12} & \cdots & \sigma_1\sigma_N\rho_{1N} \\ \sigma_2\sigma_1\rho_{21} & \sigma_2{}^2 & & \\ \cdot & & & \\ \cdot & & & \\ \cdot & & & \\ \sigma_N\sigma_1\rho_{N1} & & \cdots & \sigma_N{}^2 \end{bmatrix} \tag{5-43}$$

since $\rho_{ii} = 1$, for $i = 1, 2, \ldots, N$.   By expanding (5-43) it is easy to show that Λ_X is related to R_X by

$$\Lambda_X = R_X - \bar{X}\bar{X}^T \tag{5-44}$$

If the random process has a zero mean, then $\Lambda_X = R_X$.

The above representation for the covariance matrix is valid for both stationary and nonstationary processes.   In the case of a wide-sense stationary process, however, all the variances are the same and the correlation coefficients in a given diagonal are the same.   Thus,

$$\sigma_i{}^2 = \sigma_j{}^2 = \sigma^2 \qquad i, j = 1, 2, \ldots, N$$
$$\rho_{ij} = \rho_{|i-j|} \qquad i, j = 1, 2, \ldots, N$$

and

$$\Lambda_X = \sigma^2 \begin{bmatrix} 1 & \rho_1 & \rho_2 & \cdots & \rho_{N-1} \\ \rho_1 & 1 & \rho_1 & \cdots & \rho_{N-2} \\ \rho_2 & \rho_1 & 1 & \rho_1 & \\ \cdot & & \cdot & \cdot & \cdot \\ \cdot & & & \cdot & \cdot & \cdot \\ \cdot & & & & \cdot & \cdot & \cdot \\ & & & & & \cdot & 1 & \rho_1 \\ \rho_{N-1} & & & & & & \rho_1 & 1 \end{bmatrix} \tag{5-45}$$

Before we leave the subject of covariance matrices, it is worth noting the important role that these matrices play in connection with the joint probability density function for N random variables from a Gaussian process.   It was noted earlier that the Gaussian process was one of the few for which it is possible to write a joint probability density function for any number of random variables.   The derivation of this joint density function is beyond the scope of this discussion, but it can be shown that it becomes

$$p(x) = p[x(t_1), x(t_2), \ldots, x(t_N)]$$
$$= \frac{1}{(2\pi)^{N/2}|\Lambda_X|^{1/2}} \exp\left[-\frac{1}{2}(x^T - \bar{x}^T)\Lambda_X^{-1}(x - \bar{x})\right] \tag{5-46}$$

where $|\Lambda_X|$ is the determinant of Λ_X and Λ_X^{-1} is its inverse.

■ PROBLEMS

5-1 Consider a stationary random process having sample functions of the form shown. At periodic time instants of $t_0 \pm n t_a$ there is a rectangular pulse of width b and having a magnitude of $\pm A$ with equal probability. The time t_0 is a random variable, independent of pulse amplitude, and uniformly distributed over the period, t_a.

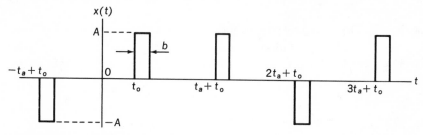

a. Find the autocorrelation function $R_X(\tau)$.
b. Find the time autocorrelation function $\mathcal{R}_x(\tau)$.
c. Can this process be ergodic?
d. What is the mean square value of the process?

5-2 A stationary random process has sample functions, $V(t)$, that are periodic rectangular waveforms as shown. The amplitude of each sample function is a random variable, X, that is uniformly distributed over the range of ± 100. The initial delay time, t_0, of each sample function is also a random variable, independent of X, that has a uniform distribution over the period, T, of the waveform. (Some simplification in calculations may result from noting that $V(t)$ can be written as $V(t) = Xf(t - t_0)$ where $f(t)$ is a unit amplitude square wave of period T.)

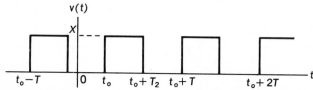

a. Compute and sketch the autocorrelation function $R_V(\tau)$.
b. Compute the time autocorrelation function $\mathcal{R}_v(\tau)$.

5-3 The random process described in Problem 5-2 has sample functions, $V(t)$, that are voltages measured across a 1-Ω resistor. Find the autocorrelation function of the power dissipated in the resistor.

5-4 A stationary random process has an autocorrelation function of the form

$$R_X(\tau) = 4\epsilon^{-|\tau|} \cos \pi\tau + \cos 3\pi\tau$$

a. What are $\overline{X^2}$ and σ^2 for the process?

b. What discrete frequency components are present?

c. What does this correlation function imply about random variables separated in time by $|\tau| = 0.25$ sec?

d. What is the shortest time separation for which the random variables $X(t)$ and $X(t + \tau)$ are uncorrelated?

e. Considering all sinusoidal components to be a signal and all other components to be noise, what is the signal-to-noise power ratio?

5-5   Consider two signals, $X(t)$ and $Y(t)$ that are sample functions from zero mean stationary random processes.   Form a new function by subtracting a constant r times the displaced waveform $Y(t + \tau)$ from $X(t)$. Thus,

$$Z(t) = X(t) - rY(t + \tau)$$

a. Find the value of r that minimizes $\sigma_Z{}^2$.

b. Over what range of values can the r determined in part (**a**) vary?

5-6   Which of the functions shown could not be an autocorrelation function?   Why?

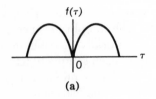

(a)

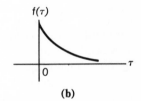

(b)

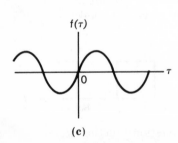

(c)

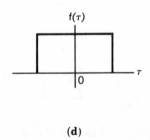

(d)

5-7   A random process has sample functions of the form

$$X(t) = Y \cos (\omega t + \theta)$$

where Y, ω, and θ are statistically independent random variables.

> Y has a mean of 2 and a variance of 4.
> θ is uniformly distributed from $-\pi$ to $+\pi$.
> ω is uniformly distributed from -5 to $+5$.

a. Is this process stationary?   Is it ergodic?

b. Find the autocorrelation function $R_X(\tau)$.

5-8 Two statistically independent random processes $X(t)$ and $Y(t)$ have autocorrelation functions given by

$$R_X(\tau) = 2e^{-2|\tau|} \cos \omega\tau$$
$$R_Y(\tau) = 9 + e^{-3|\tau^2|}$$

A third random process, $Z(t)$, has sample functions of the form

$$Z(t) = WX(t)Y(t)$$

where W is a random variable with mean 2 and variance 9.

a. What is $R_Z(\tau)$?
b. What are the mean and variance of $Z(t)$?

5-9 Two random processes have sample functions of the form

$$X(t) = A \cos (\omega_0 t + \theta) \quad \text{and} \quad Y(t) = B \sin (\omega_0 t + \theta)$$

where θ is a random variable uniformly distributed between 0 and 2π and A and B are constants.

a. What are the cross-correlation functions $R_{XY}(\tau)$ and $R_{YX}(\tau)$?
b. What is the significance of the values of these cross-correlation functions at $\tau = 0$?

5-10 Sample functions of the random process described in Problem 5-1 are applied to a half wave rectifier as shown.

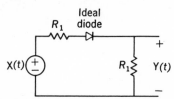

a. Find the autocorrelation function of the output $R_Y(\tau)$.
b. Find the cross-correlation function $R_{XY}(\tau)$.
c. Find the cross-correlation function $R_{YX}(\tau)$.

5-11 A random vector, $\mathbf{Z}(t)$, is composed of combinations of sample functions from two jointly stationary, statistically independent, scalar random processes, $X(t)$ and $Y(t)$.   In particular, let

$$\mathbf{Z}(t) = \begin{bmatrix} X(t) + Y(t) \\ X(t) - Y(t) \end{bmatrix}$$

If

$$R_X(\tau) = \epsilon^{-|\tau|}$$
$$R_Y(\tau) = \epsilon^{-|\tau|} + \epsilon^{-2|\tau|}$$

find the correlation matrix $\mathbf{R_z}(\tau)$.

5-12   A stationary random process having an autocorrelation function of

$$R_X(\tau) = \epsilon^{-|\tau|} \cos \pi\tau + 1$$

is sampled at periodic time instants separated by $\Delta t = 0.5$ second.

a. Write the complete covariance matrix, Λ for a set of three samples taken from this process.
b. Write the general covariance matrix for N samples.

■ REFERENCES

See References for Chapter 1, particularly Beckmann, Davenport and Root, Papoulis, and Thomas.

Spectral Density

6-1 Introduction

The use of Fourier transforms and Laplace transforms in the analysis of linear systems is widespread and frequently leads to much saving in labor. The principal reason for this simplification is that the convolution integral of time-domain methods is replaced by simple multiplication when frequency-domain methods are used.

In view of this widespread use of frequency-domain methods it is natural to ask if such methods are still useful when the inputs to the system are random. The answer to this question is that they *are* still useful but that some modifications are required and that a little more care is necessary in order to avoid pitfalls. However, when properly used, frequency-domain methods offer essentially the same advantages in dealing with random signals as they do with nonrandom signals.

Before beginning this discussion, it is desirable to review briefly the frequency-domain representation of a nonrandom time function. The most natural representation of this sort is the Fourier transform, which leads to the concept of frequency spectrum. Thus, the Fourier transform of some nonrandom time function, $f(t)$, is defined to be

$$F(\omega) = \int_{-\infty}^{\infty} f(t)\epsilon^{-j\omega t}\, dt \qquad \textbf{(6-1)}$$

If $f(t)$ is a voltage, say, then $F(\omega)$ has the units of volts per hertz and represents the *relative* magnitude and phase of steady-state sinusoids (of frequency ω) that can be summed to produce the original $f(t)$. Thus, the magnitude of the Fourier transform has the physical significance of being the *amplitude density* as a function of frequency and as such gives a clear indication of how the energy of $f(t)$ is distributed with respect to frequency.

It might seem reasonable to use exactly the same procedure in dealing with *random* signals—that is, to use the Fourier transform of any particular sample function $x(t)$, defined by

$$F_x(\omega) = \int_{-\infty}^{\infty} x(t)\epsilon^{-j\omega t}\, dt$$

as the frequency-domain representation of the random process. This is not possible, however, for at least two reasons. In the first place, the Fourier transform will be a random variable over the ensemble (for any fixed ω) since it will have a different value for each member of the ensemble of possible sample functions. Hence, it cannot be a frequency representation of the process, but only of one member of the process. However, it might still be possible to use this function by finding its expected value (or mean) over the ensemble if it were not for the second reason. The second, and more basic, reason for not using the $F_x(\omega)$ just defined is that—for stationary processes, at least—it almost never exists! It may be recalled that one of the conditions for a time function to be Fourier transformable is that it be absolutely integrable; that is,

$$\int_{-\infty}^{\infty} |x(t)|\, dt < \infty \tag{6-2}$$

This condition can never be satisfied by any non-zero sample function from a wide-sense stationary random process. The Fourier transform in the ordinary sense will never exist in this case, although it may occasionally exist in the sense of generalized functions, including impulses, and so forth.

Now that the usual Fourier transform has been ruled out as a means of obtaining a frequency-domain representation for a random process, the next thought is to use the Laplace transform, since this contains a built-in convergence factor. Of course, the usual one-sided transform, which considers $f(t)$ for $t \geq 0$ only, is not applicable for a wide-sense stationary process; however, this is no real difficulty since the two-sided Laplace transform is good for negative values of time as well as positive. Once this is done, the Laplace transform for almost any sample function from a stationary random process will exist.

It turns out, however, that this approach is not so promising as it looks since it merely transfers the existence problems from the transform to the inverse transform. A study of these problems requires a knowledge of complex variable theory that is beyond the scope of the present discussion.   Hence, it appears that the simplest mathematically-accept-able approach is to return to the Fourier transform and employ an artifice that will insure existence.   Even in this case it will not be possible to justify rigorously all the steps, and a certain amount of the procedure will have to be accepted on faith.

6-2   Relation of Spectral Density to the Fourier Transform

In order to use the Fourier transform technique it is necessary to modify the sample functions of a stationary random process in such a way that the transform of each sample function exists.   There are many ways in which this might be done, but the simplest one is to define a new sample function having finite duration.   Thus, let

$$
\begin{aligned}
X_T(t) &= X(t) &\quad |t| \le T < \infty \\
&= 0 &\quad |t| > T
\end{aligned}
\tag{6-3}
$$

and note that the truncated time function $X_T(t)$ will satisfy the condition of (6-2), as long as T remains finite, provided that the stationary process from which it is taken has a finite mean-square value.   Hence, $X_T(t)$ will be Fourier transformable.   In fact, $X_T(t)$ will satisfy the more stringent requirement for integrable square functions; that is

$$
\int_{-\infty}^{\infty} |X_T(t)|^2 \, dt \quad < \infty
\tag{6-4}
$$

This condition will be needed in the subsequent development.

Since $X_T(t)$ is Fourier transformable, its transform may be written as

$$
F_X(\omega) = \int_{-\infty}^{\infty} X_T(t)\epsilon^{-j\omega t} \, dt \quad T < \infty
\tag{6-5}
$$

Eventually it will be necessary to let T increase without limit; the purpose of the following discussion is to show that the *expected value* of $|F_X(\omega)|^2$ does exist in the limit even though the $F_X(\omega)$ for any one sample function does not.   The first step in demonstrating this is to apply Parseval's theorem to $X_T(t)$ and $F_X(\omega)$.[1]   Thus, since $x_T(t) = 0$ for $|t| > T$,

$$
\int_{-T}^{T} X_T^2(t) \, dt = \frac{1}{2\pi} \int_{-\infty}^{\infty} |F_X(\omega)|^2 \, d\omega
\tag{6-6}
$$

Note that $|F_X(\omega)|^2 = F_X(\omega)F_X(-\omega)$ since $F_X(-\omega)$ is the complex conjugate of $F_X(\omega)$ when $X_T(t)$ is a real time function.

Since the quantity being sought is the distribution of average power

---

[1] Parseval's theorem states that if $f(t)$ and $g(t)$ are transformable time functions with transforms of $F(\omega)$ and $G(\omega)$ respectively, then

$$
\int_{-\infty}^{\infty} f(t)g(t) \, dt = \frac{1}{2\pi} \int_{-\infty}^{\infty} F(\omega)G(-\omega) \, d\omega
$$

as a function of frequency, the next step is to average both sides of (6-6) over the total time, $2T$. Hence, dividing both sides by $2T$ gives

$$\frac{1}{2T} \int_{-T}^{T} X_T^2(t) \, dt = \frac{1}{4\pi T} \int_{-\infty}^{\infty} |F_X(\omega)|^2 \, d\omega \qquad (6\text{-}7)$$

The left side of (6-7) is seen to be proportional to the average power of the sample function in the time interval from $-T$ to T. More exactly, it is the square of the *effective* value of $X_T(t)$. Furthermore, for an ergodic process, this quantity would approach the mean-square value of the process as T approached infinity.

However, it is not possible at this stage to let T approach infinity, since $F_X(\omega)$ simply does not exist in the limit. It should be remembered, though, that $F_X(\omega)$ is a random variable with respect to the ensemble of sample functions from which $X(t)$ was taken. It is reasonable to suppose (and can be rigorously proved) that the limit of the *expected value* of $(1/T)|F_X(\omega)|^2$ does exist since the integral of this "always positive" quantity certainly does exist, as shown by (6-4). Hence, taking the expectation of both sides of (6-7), interchanging the expectation and integration and then taking the limit as $T \rightarrow \infty$ we obtain

$$E\left\{\frac{1}{2T} \int_{-T}^{T} X_T^2(t) \, dt\right\} = E\left\{\frac{1}{4\pi T} \int_{-\infty}^{\infty} |F_X(\omega)|^2 \, d\omega\right\}$$

$$\lim_{T \to \infty} \frac{1}{2T} \int_{-T}^{T} \overline{X^2} \, dt = \lim_{T \to \infty} \frac{1}{4\pi T} \int_{-\infty}^{\infty} E\{|F_X(\omega)|^2\} \, d\omega \qquad (6\text{-}8)$$

$$\langle \overline{X^2} \rangle = \frac{1}{2\pi} \int_{-\infty}^{\infty} \lim_{T \to \infty} \frac{E\{|F_X(\omega)|^2\}}{2T} \, d\omega$$

For a stationary process, the time average of the mean-square value is equal to the mean-square value and (6-8) can be written as

$$\overline{X^2} = \frac{1}{2\pi} \int_{-\infty}^{\infty} \lim_{T \to \infty} \frac{E\{|F_X(\omega)|^2\}}{2T} \, d\omega \qquad (6\text{-}9)$$

The integrand of the right side of (6-9), which will be designated by the symbol $S_X(\omega)$, is called the *spectral density* of the random process. Thus,

$$S_X(\omega) = \lim_{T \to \infty} \frac{E[|F_X(\omega)|^2]}{2T} \qquad (6\text{-}10)$$

and it must be remembered that it is not possible to let $T \rightarrow \infty$ *before* taking the expectation. If $X(t)$ is a voltage, say, then $S_X(\omega)$ has the units of volts² per hertz, and its integral, as shown by (6-9), leads to the mean-square value; that is,

$$\overline{X^2} = \frac{1}{2\pi} \int_{-\infty}^{\infty} S_X(\omega) \, d\omega \qquad (6\text{-}11)$$

The physical interpretation of spectral density can be made somewhat clearer by thinking in terms of average power, although this is a fairly specialized way of looking at it. If $X(t)$ is a voltage or current associated with a 1 Ω resistance, then $\overline{X^2}$ is just the average power dissipated in that resistance. The spectral density, $S_X(\omega)$, can then be interpreted as the average power associated with a bandwidth of 1 Hz centered at $\omega/2\pi$ Hz. [Note that the unit of bandwidth is the hertz (or cycle per second) and not the radian per second, because of the factor of $1/(2\pi)$ in the integral of (6-11).]

The foregoing analysis of spectral density has been carried out in somewhat more detail than is customary in an introductory discussion. The reason for this is an attempt to avoid some of the mathematical pitfalls that a more superficial approach might gloss over. There is no doubt that this method makes the initial study of spectral density more difficult for the reader, but it is felt that the additional rigor is well worth the effort. Furthermore, even if all of the implications of the discussion are not fully understood, it should serve to make the reader aware of the existence of some of the less obvious difficulties of frequency-domain methods.

Another approach to spectral density, which treats it as a defined quantity based on the autocorrelation function, is given in Section 6-6. From the standpoint of application, such a definition is probably more useful than the more basic approach given here and is also easier to understand. It does not, however, make the physical interpretation as apparent as the basic derivation does.

Before turning to a more detailed discussion of the properties of spectral densities, it may be noted that in system analysis the spectral density of the input random process will play the same role as does the transform of the input in the case of nonrandom signals. The major difference is that spectral density represents a *power* density rather than a *voltage* density. Thus, it will be necessary to define a *power transfer function* for the system rather than a *voltage transfer function*.

Exercise 6-2

The spectral density defined above is sometimes referred to as the "two-sided spectral density" since it exists for both positive and negative values of ω. However, some authors prefer to define a "one-sided spectral density," which is expressed as a function of $f = \omega/2\pi$ and exists only for positive values of f. If this one-sided spectral density is designated as $G_X(f)$, then the mean-square value of the random process is given by

$$\overline{X^2} = \int_0^\infty G_X(f)\, df$$

What is the relationship between $G_X(f)$ and $S_X(\omega)$ for any value of $\omega = 2\pi f$?

Answer: $G_X(f) = 2S_X(2\pi f)$ $\quad f \geq 0$
$= 0$ $\quad\quad\quad f < 0$

6-3 Properties of Spectral Density

Most of the important properties of spectral density are summarized by the simple statement that it is a *real, positive, even* function of ω. It is known from the study of Fourier transforms that their *magnitude* is certainly real and positive. Hence, the expected value will also possess the same properties.

A special class of spectral densities, which is more commonly used than any other, is said to be *rational*, since it is composed of a ratio of polynomials. Since the spectral density is an even function of ω, these polynomials involve only even powers of ω. Thus, it is represented by

$$S_X(\omega) = \frac{S_0(\omega^{2n} + a_{2n-2}\omega^{2n-2} + \cdots + a_2\omega^2 + a_0)}{\omega^{2m} + b_{2m-2}\omega^{2m-2} + \cdots + b_2\omega^2 + b_0} \tag{6-12}$$

If the mean-square value of the random process is finite, then the area under $S_X(\omega)$ must also be finite, from (6-11). In this case it is necessary that $m > n$. This condition will always be assumed here except for a very special case of *white noise*. White noise is a term applied to a random process for which the spectral density is constant for all ω; that is, $S_X(\omega) = S_0$. Although such a process cannot exist physically (since it has infinite mean-square value), it is a convenient mathematical fiction, which greatly simplifies many computations that would otherwise be very difficult. The justification and illustration of the use of this concept will be discussed in more detail later.

Spectral densities of this type are continuous and, as such, cannot represent random processes having dc or periodic components. The reason is not difficult to understand when spectral density is interpreted as average power per unit bandwidth. Any dc component in a random process represents a finite average power in *zero* bandwidth, since this component has a discrete frequency spectrum. Finite power in zero bandwidth is equivalent to an infinite power density. Hence, we would expect the spectral density in this case to be infinite at zero frequency but finite elsewhere; that is, it would contain a δ function at $\omega = 0$. A similar argument for periodic components would justify the existence of δ functions at these discrete frequencies. A rigorous derivation of these results will serve to make the argument more precise and, at the same time, illustrate the use of the defining equation, (6-10), in the calculation of spectral densities.

In order to carry out the desired derivation, consider a stationary random process having sample functions of the form

$$X(t) = A + B \cos(\omega_1 t + \theta) \tag{6-13}$$

where A, B, and ω_1 are constants and θ is a random variable uniformly distributed from 0 to 2π; that is,

$$p(\theta) = \frac{1}{2\pi} \qquad 0 \le \theta \le 2\pi$$

$$0 \qquad \text{elsewhere}$$

The Fourier transform of the truncated sample function, $X_T(t)$, is

$$F_X(\omega) = \int_{-T}^{T} [A + B \cos (\omega_1 t + \theta)]\epsilon^{-j\omega t} dt$$

$$= A \left. \frac{\epsilon^{-j\omega t}}{-j\omega} \right|_{-T}^{T} + \frac{B}{2} \left. \frac{\epsilon^{j[(\omega_1-\omega)t+\theta]}}{j(\omega_1 - \omega)} \right|_{-T}^{T} + \frac{B}{2} \left. \frac{\epsilon^{-j[(\omega_1+\omega)t+\theta]}}{-j(\omega_1 + \omega)} \right|_{-T}^{T}$$

Substituting in the limits and simplifying leads immediately to

$$F_X(\omega) = \frac{2A \sin \omega T}{\omega} + B \left[\frac{\epsilon^{j\theta} \sin (\omega - \omega_1)T}{(\omega - \omega_1)} + \frac{\epsilon^{-j\theta} \sin (\omega + \omega_1)T}{(\omega + \omega_1)} \right]$$

$$(6\text{-}14)$$

The square of the magnitude of $F_X(\omega)$ will have nine terms, some of which are independent of the random variable θ and the rest of which involve either $\epsilon^{\pm j\theta}$ or $\epsilon^{\pm j2\theta}$. In anticipation of the result that the expectation of all terms involving θ will vanish, it is convenient to write the squared magnitude in symbolic form without bothering to determine all the coefficients. Thus,

$$|F_X(\omega)|^2 = \frac{4A^2 \sin^2 \omega T}{\omega^2} + B^2 \left[\frac{\sin^2 (\omega - \omega_1)T}{(\omega - \omega_1)^2} + \frac{\sin^2 (\omega + \omega_1)T}{(\omega + \omega_1)^2} \right]$$
$$+ C(\omega)\epsilon^{j\theta} + C(-\omega)\epsilon^{-j\theta} + D(\omega)\epsilon^{j2\theta} + D(-\omega)\epsilon^{-j2\theta} \quad (6\text{-}15)$$

Now consider the expected value of any term involving θ. These are all of the form $G(\omega)\epsilon^{jn\theta}$, and the expected value is

$$E[G(\omega)\epsilon^{jn\theta}] = G(\omega) \int_{0}^{2\pi} \frac{1}{2\pi} \epsilon^{jn\theta} d\theta = \left. \frac{G(\omega)}{2\pi} \frac{\epsilon^{jn\theta}}{jn} \right|_{0}^{2\pi} \qquad (6\text{-}16)$$

$$= 0 \qquad n = \pm 1, \pm 2, \ldots$$

Thus, the last four terms of (6-15) will vanish and the expected value will become

$$E[|F_X(\omega)|^2] = 4A^2 \left[\frac{\sin^2 \omega T}{\omega^2} \right] + B^2 \left[\frac{\sin^2 (\omega - \omega_1)T}{(\omega - \omega_1)^2} + \frac{\sin^2 (\omega + \omega_1)T}{(\omega + \omega_1)^2} \right]$$

$$(6\text{-}17)$$

From (6-10), the spectral density is

$$S_X(\omega) = \lim_{T \to \infty} \left\{ 2A^2 T \left[\frac{\sin \omega T}{\omega T} \right]^2 + \frac{B^2 T}{2} \left[\frac{\sin (\omega - \omega_1)T}{(\omega - \omega_1)T} \right]^2 \right.$$

$$\left. + \frac{B^2 T}{2} \left[\frac{\sin (\omega + \omega_1)T}{(\omega + \omega_1)T} \right]^2 \right\}$$

$$(6\text{-}18)$$

In order to investigate the limit, consider the essential part of the first term; that is,

$$\lim_{T \to \infty} T \left[\frac{\sin \omega T}{\omega T} \right]^2 = ?$$

This limit is clearly zero when ω is *not* zero since $\sin^2 \omega T$ cannot exceed 1 and the denominator increases as T. When $\omega = 0$, however,

$$\frac{\sin \omega T}{\omega T} \bigg|_{\omega = 0} = 1$$

and the limit is infinite. Hence, one can write

$$\lim_{T \to \infty} T \left[\frac{\sin \omega T}{\omega T} \right]^2 = K\delta(\omega) \tag{6-19}$$

where K represents the area of the δ function and has not yet been evaluated. The value of K can be found by equating the areas of both sides of (6-19); that is,

$$\lim_{T \to \infty} \int_{-\infty}^{\infty} T \left[\frac{\sin \omega T}{\omega T} \right]^2 d\omega = \int_{-\infty}^{\infty} K\delta(\omega) \, d\omega \tag{6-20}$$

The integral on the left is tabulated and has a value of π for *all* values of $T > 0$. Thus, the limiting operation becomes trivial, and (6-20) leads to

$$\pi = K$$

An exactly similar procedure can be used for the other terms in (6-18). It is left as an exercise for the reader to show that the final result becomes

$$S_X(\omega) = 2\pi A^2 \delta(\omega) + \frac{\pi}{2} B^2 \delta(\omega - \omega_1) + \frac{\pi}{2} B^2 \delta(\omega + \omega_1) \tag{6-21}$$

This spectral density is shown in Figure 6-1.

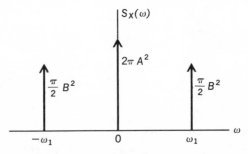

Figure 6-1 Spectral density of dc and sinusoidal components.

It is of interest to determine the area of the spectral density in order to verify that (6-21) does in fact lead to the proper mean-square value.

Thus, according to (6-11)

$$\overline{X^2} = \frac{1}{2\pi} \int_{-\infty}^{\infty} \left[2\pi A^2 \delta(\omega) + \frac{\pi}{2} B^2 \delta(\omega - \omega_1) + \frac{\pi}{2} B^2 \delta(\omega + \omega_1) \right] d\omega$$

$$= \frac{1}{2\pi} \left[2\pi A^2 + \frac{\pi}{2} B^2 + \frac{\pi}{2} B^2 \right] = A^2 + \frac{1}{2} B^2$$

(6-22)

The reader can verify easily that this same result would be obtained from the ensemble average of $X^2(t)$.

It is also possible to have spectral densities with both a continuous component and discrete components. An example of this sort that arises frequently in connection with communication systems or sampled data control systems is the random amplitude pulse sequence shown in Figure 6-2. It is assumed here that all of the pulses have the same shape

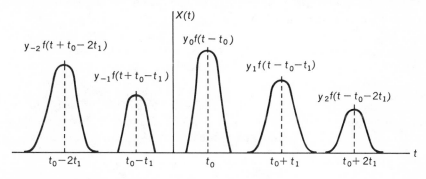

Figure 6-2 Random amplitude pulse sequence.

but their amplitudes are random variables that are statistically independent from pulse to pulse. However, all the amplitude variables have the same mean, $\overline{Y}$, and the same variance, σ_Y^2. The repetition period for the pulses is t_1, a constant, and the reference time for any sample function is t_0, which is a random variable uniformly distributed over an interval of t_1.

The complete derivation of the spectral density is too lengthy to be included here, but the final result indicates some interesting points. This result may be expressed in terms of the Fourier transform $F(\omega)$ of the basic pulse shape $f(t)$, and is

$$S_X(\omega) = |F(\omega)|^2 \left[\frac{\sigma_Y^2}{t_1} + \frac{2\pi(\overline{Y})^2}{t_1^2} \sum_{n=-\infty}^{\infty} \delta\left(\omega - \frac{2\pi n}{t_1}\right) \right]$$

(6-23)

If the basic pulse shape is rectangular, with a width of t_2, the corresponding spectral density will be as shown in Figure 6-3. From (6-23) the following general conclusions are possible:

1. Both the continuous spectrum amplitude and the areas of the

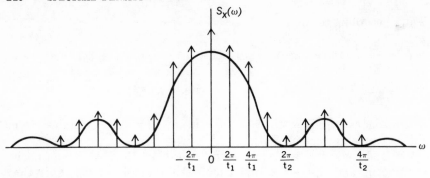

Figure 6-3 Spectral density for rectangular pulse sequence with random amplitudes.

δ functions are proportional to the squared magnitude of the Fourier transform of the basic pulse shape.

2. If the mean value of the pulse amplitude is zero, there will be *no discrete spectrum* even though the pulses occur periodically.

3. If the variance of the pulse amplitude is zero, there will be *no continuous spectrum*.

Another property of spectral densities concerns the derivative of the random process. Suppose that $\dot{X}(t) = dX(t)/dt$ and that $X(t)$ has a spectral density of $S_X(\omega)$, which was defined as

$$S_X(\omega) = \lim_{T \to \infty} \frac{E[|F_X(\omega)|^2]}{2T}$$

The truncated version of the derivative, $\dot{X}_T(t)$, will have a Fourier transform of $j\omega F_X(\omega)$, with the possible addition of two constant terms (arising from the discontinuities at $\pm T$) that will vanish in the limit. Hence, the spectral density of the derivative becomes

$$
\begin{aligned}
S_{\dot{X}}(\omega) &= \lim_{T \to \infty} \frac{E[|j\omega F_X(\omega)(-j\omega)F_X(-\omega)|]}{2T} \\
&= \omega^2 \lim_{T \to \infty} \frac{E[|F_X(\omega)|^2]}{2T} = \omega^2 S_X(\omega)
\end{aligned}
$$

(6-24)

It is seen, therefore, that differentiation creates a new process whose spectral density is simply ω^2 times the spectral density of the original process. In this connection, it should be noted that if $S_X(\omega)$ is finite at $\omega = 0$, then $S_{\dot{X}}(\omega)$ will be zero at $\omega = 0$. Furthermore, if $S_X(\omega)$ does not drop off more rapidly than $1/\omega^2$ as $\omega \to \infty$, then $S_{\dot{X}}(\omega)$ will approach a constant at large ω and the mean-square value for the derivative will be infinite. This corresponds to the case of nondifferentiable random processes.

Exercise 6-3

A sample function from a stationary random process consists of a repetitive random amplitude binary waveform (amplitudes of 0 or 1 with equal probability) having a period of 1 mse.   Evaluate the spectral density at $f = 0, 500$, and 1000 Hz.

Answer: $0, 1.01 \times 10^{-4}, 2.5 \times 10^{-4} + .25\delta(f)$

6-4   Spectral Density and the Complex Frequency Plane

In the discussion so far the spectral density has been expressed as a function of the real angular frequency ω.   However, for applications to system analysis, it is very convenient to express it in terms of the complex frequency s, since system transfer functions are more convenient in this form.   This change can be made very simply by replacing $j\omega$ with s.   Hence, along the $j\omega$-axis of the complex frequency plane the spectral density will be the same as that already discussed.

The formal conversion to complex frequency representation is accomplished by replacing ω by $-js$ or ω^2 by $-s^2$.   The resulting spectral density should properly be designated as $S_X(-js)$, but this notation is somewhat clumsy.   Therefore, spectral density in the *s-plane* will be designated simply as $S_X(s)$.   It is evident that $S_X(s)$ and $S_X(\omega)$ are somewhat different functions of their respective arguments, so that the notation is symbolic rather than precise.

For the special case of rational spectral densities, in which only even powers of ω occur, this substitution is equivalent to replacing ω^2 by $-s^2$.   For example, consider the rational spectrum

$$S_X(\omega) = \frac{10(\omega^2 + 5)}{\omega^4 + 10\omega^2 + 24}$$

When expressed as a function of s, this becomes

$$S_X(s) = S_X(-j\omega) = \frac{10(-s^2 + 5)}{s^4 - 10s^2 + 24} \tag{6-25}$$

Any spectral density can also be represented (except for a constant of proportionality) in terms of its pole-zero configuration in the complex frequency plane.   Such a representation is often convenient in carrying out certain calculations, which will be discussed in the following sections.   For purposes of illustration, consider the spectral density of (6-25).   This may be factored as

$$S_X(s) = \frac{-10(s + \sqrt{5})(s - \sqrt{5})}{(s + 2)(s - 2)(s + \sqrt{6})(s - \sqrt{6})}$$

and the pole-zero configuration plotted as shown in Figure 6-4.   This plot also illustrates the important point that such configurations are always symmetrical about the $j\omega$ axis.   When the spectral density is not rational, the substitution is the same but may not be quite as straight-

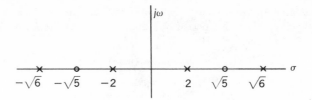

Figure 6-4   Pole-zero configuration for a spectral density.

forward. For example, the spectral density given by (6-23) could be expressed in the complex frequency plane as

$$S_X(s) = F(s)F(-s)\left[\frac{\sigma_Y^2}{t_1} + \frac{2\pi(\bar{Y})^2}{t_1^2}\sum_{n=-\infty}^{\infty}\delta\left(s - j\frac{2\pi n}{t_1}\right)\right] \qquad \text{(6-26)}$$

where $F(s)$ is the *Laplace* transform of the basic pulse shape $f(t)$.

In addition to making spectral densities more convenient for system analysis, the use of the complex frequency s also makes it more convenient to evaluate mean-square values. This application is discussed in the following section.

Exercise 6-4

A spectral density has the form

$$S_X(\omega) = \frac{1}{\omega^4 + 4}$$

Find the pole locations for this spectral density in the complex frequency plane.

Answer: $s = \pm 1 \pm j1$

6-5   Mean-Square Values from Spectral Density

It was shown in the course of defining the spectral density that the mean-square value of the random process was given by

$$\overline{X^2} = \frac{1}{2\pi}\int_{-\infty}^{\infty} S_X(\omega)\,d\omega \qquad \text{(6-11)}$$

Hence, the mean-square value is proportional to the *area* of the spectral density.

The evaluation of an integral such as (6-11) may be very difficult if the spectral density has a complicated form or if it involves high powers of ω. A classical way of carrying out such integration is to convert the variable of integration to a complex variable (by substituting s for $j\omega$) and then to utilize some powerful theorems concerning integration around closed paths in the complex plane. This is probably the easiest and most satisfactory way of obtaining mean-square values but, unfortunately, requires a knowledge of complex variables that the reader may not possess. The *mechanics* of the procedure will be discussed at the end of this section, however, for those interested in this method.

An alternative method, which will be discussed first, is to utilize some tabulated results for spectral densities that are rational. These have been tabulated in general form for polynomials of various degrees and their use is simply a matter of substituting in the appropriate numbers. The existence of such general forms is primarily a consequence of the symmetry of the spectral density. As a result of this symmetry, it is always possible to factor rational spectral densities into the form

$$S_X(s) = \frac{c(s)c(-s)}{d(s)\,d(-s)} \tag{6-27}$$

where $c(s)$ contains the left-half-plane (lhp) zeros, $c(-s)$ the right-half-plane (rhp) zeros, $d(s)$ the lhp poles, and $d(-s)$ the rhp poles.

When the real integration of (6-11) is expressed in terms of the complex variable s, the mean-square value becomes

$$\overline{X^2} = \frac{1}{2\pi j} \int_{-j\infty}^{j\infty} S_X(s)\,ds = \frac{1}{2\pi j} \int_{-j\infty}^{j\infty} \frac{c(s)c(-s)}{d(s)\,d(-s)}\,ds \tag{6-28}$$

For the special case of rational spectral densities, $c(s)$ and $d(s)$ are polynomials in s and may be written as

$$c(s) = c_{n-1}s^{n-1} + c_{n-2}s^{n-2} + \cdots + c_0$$
$$d(s) = d_n s^n + d_{n-1}s^{n-1} + \cdots + d_0$$

Some of the coefficients of $c(s)$ may be zero, but $d(s)$ must be of higher degree than $c(s)$ and must not have any coefficients missing.

Integrals of the form in (6-28) have been tabulated for values of n up to 10, although beyond $n = 3$ or 4 the general results are so complicated as to be of doubtful value. An abbreviated table is given in Table 6-1.

Table 6-1 TABLE OF INTEGRALS

$$I_n = \frac{1}{2\pi j} \int_{-j\infty}^{j\infty} \frac{c(s)c(-s)}{d(s)\,d(-s)}\,ds$$
$$c(s) = c_{n-1}s^{n-1} + c_{n-2}s^{n-2} + \cdots + c_0$$
$$d(s) = d_n s^n + d_{n-1}s^{n-1} + \cdots + d_0$$

$$I_1 = \frac{c_0^2}{2d_0 d_1}$$

$$I_2 = \frac{c_1^2 d_0 + c_0^2 d_2}{2d_0 d_1 d_2}$$

$$I_3 = \frac{c_2^2 d_0 d_1 + (c_1^2 - 2c_0 c_2)d_0 d_3 + c_0^2 d_2 d_3}{2d_0 d_3 (d_1 d_2 - d_0 d_3)}$$

As an example of this calculation, consider the spectral density

$$S_X(\omega) = \frac{\omega^2 + 4}{\omega^4 + 10\omega^2 + 9}$$

When ω is replaced by $-js$, this becomes

$$S_X(s) = \frac{-(s^2 - 4)}{s^4 - 10s^2 + 9} = \frac{-(s^2 - 4)}{(s^2 - 1)(s^2 - 9)} \qquad \text{(6-29)}$$

This can be factored into

$$S_X(s) = \frac{(s + 2)(-s + 2)}{(s + 1)(s + 3)(-s + 1)(-s + 3)} \qquad \text{(6-30)}$$

from which it is seen that

$$c(s) = s + 2$$
$$d(s) = (s + 1)(s + 3) = s^2 + 4s + 3$$

This is a case in which $n = 2$ and

$$c_1 = 1$$
$$c_0 = 2$$
$$d_2 = 1$$
$$d_1 = 4$$
$$d_0 = 3$$

From Table 6-1, I_2 is given by

$$I_2 = \frac{c_1^2 d_0 + c_0^2 d_2}{2d_0 d_1 d_2} = \frac{(1)^2(3) + (2)^2(1)}{2(3)(4)(1)} = \frac{3 + 4}{24} = \frac{7}{24}$$

However, $\overline{X^2} = I_2$, so that

$$\overline{X^2} = \frac{7}{24}$$

The procedure just presented is a mechanical one and in order to be a useful tool does not require any deep understanding of the theory. Some precautions are necessary, however. In the first place, as noted above, it is necessary that $c(s)$ be of lower degree than $d(s)$. Second, it is necessary that $c(s)$ and $d(s)$ have roots *only* in the left half plane. Finally, it is necessary that $d(s)$ have *no* roots *on* the $j\omega$-axis.

It was noted previously that the use of complex integration provides a very general and very powerful method of evaluating integrals of the form given in (6-28). A brief summary of the theory of such integration is given in Appendix F, and these ideas will be utilized here to demonstrate another method of evaluating mean-square values from spectral density. As a means of acquainting the student with the potential usefulness of this general procedure, only the mechanics of this method will be discussed. The student should be aware, however, that there are many pitfalls associated with using mathematical tools without having a thorough grasp of their theory and, hence, he is encouraged to acquire the proper theoretical understanding as soon as possible.

The latter method is based on the evaluation of residues, in much the same way as is done in connection with finding inverse Laplace transforms. Consider, for example, the spectral density given above in (6-29) and (6-30). This spectral density may be represented by the pole-zero configuration shown in Figure 6-5. The path of integration called

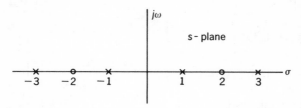

Figure 6-5   Pole-zero configuration for a spectral density.

for by (6-28) is along the $j\omega$-axis, but the methods of complex integration discussed in Appendix F require a closed path. Such a closed path can be obtained by adding a semicircle at infinity that encloses either the left half plane or the right half plane. Less difficulty with the algebraic signs is encountered if the left half plane is used, so the path shown in Figure 6-6 will be assumed from now on. In order for the integral

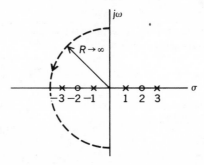

Figure 6-6   Path of integration for evaluating mean-square value.

around this closed path to the same as the integral along the $j\omega$ axis, it is necessary for the contribution due to the semicircle to vanish as $R \to \infty$. For rational spectral densities this will be true whenever the denominator polynomial is of higher degree than the numerator polynomial (since only even powers are present).

A basic result of complex variable theory states that the value of an integral around a simple closed contour in the complex plane is equal to $2\pi j$ times the sum of the residues at the poles contained *within* that contour (see (F-3), Appendix F). Since the expression for the mean-square value has a factor of $1/(2\pi j)$, and since the chosen contour completely encloses the left half plane, it follows that the mean-square value can be expressed in general as

$$\overline{X^2} = \Sigma \text{ (residues at lhp poles)} \tag{6-31}$$

For the example being considered, the only lhp poles are at -1 and -3. The residues can be evaluated easily by multiplying $S_X(s)$ by the factor containing the pole in question and letting s assume the value of the pole. Thus,

$$K_{-1} = [(s + 1)S_X(s)]_{s=-1} = \left[\frac{-(s + 2)(s - 2)}{(s - 1)(s + 3)(s - 3)}\right]_{s=-1} = \frac{3}{16}$$

$$K_{-3} = [(s + 3)S_X(s)]_{s=-3} = \left[\frac{-(s + 2)(s - 2)}{(s + 1)(s - 1)(s - 3)}\right]_{s=-3} = \frac{5}{48}$$

From (6-31) it follows that

$$\overline{X^2} = \frac{3}{16} + \frac{5}{48} = \frac{7}{24}$$

which is the same value obtained above.

If the poles are not simple, the more general procedures discussed in Appendix F may be employed for evaluating the residues. However, the mean-square value is still obtained from (6-31).

Exercise 6-5

Using contour integration, determine the mean-square value of the random process whose spectral density is

$$S_X(\omega) = \frac{\omega^2}{\omega^4 + 3\omega^2 + 2}$$

Check this answer using Table 6-1.

Answer: $1/\sqrt{2} - 1/2$

6-6   Relation of Spectral Density to the Autocorrelation Function

The autocorrelation function was shown in Chapter 5 to be the expected value of the product of time functions. In this chapter, it has been shown that the spectral density is related to the expected value of the product of Fourier transforms. It would appear, therefore, that there should be some direct relationship between these two expected values. Almost intuitively one would expect the spectral density to be the Fourier (or Laplace) transform of the autocorrelation function, and this turns out to be the case.

We will consider first the case of a nonstationary random process and then specialize the result to a stationary process. In (6-10) the spectral density was defined as

$$S_X(\omega) = \lim_{T \to \infty} \frac{E[|F_X(\omega)|^2]}{2T} \tag{6-10}$$

where $F_X(\omega)$ is the Fourier transform of the truncated sample function. Thus,

$$F_X(\omega) = \int_{-T}^{T} X_T(t)\epsilon^{-j\omega t} \, dt \qquad T < \infty \tag{6-32}$$

Substituting (6-32) into (6-10) yields

$$S_X(\omega) = \lim_{T \to \infty} \frac{1}{2T} E\left[\int_{-T}^{T} X_T(t_1)\epsilon^{+j\omega t_1} \, dt_1 \int_{-T}^{T} X_T(t_2)\epsilon^{-j\omega t_2} \, dt_2 \right] \tag{6-33}$$

since $|F_X(\omega)|^2 = F_X(\omega)F_X(-\omega)$. The subscripts on t_1 and t_2 have been introduced so that we can distinguish the variables of integration when the *product* of integrals is rewritten as an *iterated double integral*. Thus, write (6-33) as

$$S_X(\omega) = \lim_{T \to \infty} \frac{1}{2T} E\left[\int_{-T}^{T} dt_2 \int_{-T}^{T} \epsilon^{-j\omega(t_2-t_1)} X_T(t_1) X_T(t_2) \, dt_1 \right]$$

$$= \lim_{T \to \infty} \frac{1}{2T} \int_{-T}^{T} dt_2 \int_{-T}^{T} \epsilon^{-j\omega(t_2-t_1)} E[X_T(t_1) X_T(t_2)] \, dt_1 \tag{6-34}$$

Moving the expectation operation inside the double integral can be shown to be valid in this case, but the details will not be discussed here.

The expectation in the integrand above is recognized as the autocorrelation function of the truncated process. Thus,

$$E[X_T(t_1)X_T(t_2)] = R_X(t_1,t_2) \qquad |t_1|, |t_2| \leq T$$

$$= 0 \qquad \text{elsewhere} \tag{6-35}$$

Making the substitution

$$t_2 - t_1 = \tau$$
$$dt_2 = d\tau$$

we can write (6-35) as

$$S_X(\omega) = \lim_{T \to \infty} \frac{1}{2T} \int_{-T-t_1}^{T-t_1} d\tau \int_{-T}^{T} \epsilon^{-j\omega\tau} R_X(t_1, t_1 + \tau) \, dt_1$$

when the limits on t_1 are imposed by (6-35). Interchanging the order of integration and moving the limit inside the τ-integral gives

$$S_X(\omega) = \int_{-\infty}^{\infty} \left\{ \lim_{T \to \infty} \frac{1}{2T} \int_{-T}^{T} R_X(t_1, t_1 + \tau) \, dt_1 \right\} \epsilon^{-j\omega\tau} \, d\tau \tag{6-36}$$

From (6-36) it is apparent that the spectral density is the Fourier transform of the time average of the autocorrelation function. This may be expressed in shorter notation as follows:

$$S_X(\omega) = \mathcal{F}\{\langle R_X(t, t + \tau)\rangle\} \tag{6-37}$$

The relationship given in (6-37) is valid for nonstationary processes also.

If the process in question is a stationary random process, the auto-correlation function is independent of time; therefore,

$$\langle R_X(t_1, t_1 + \tau)\rangle = R_X(\tau)$$

Accordingly, the spectral density of a wide-sense stationary random process is just the Fourier transform of the autocorrelation function; that is,

$$\begin{aligned} S_X(\omega) &= \int_{-\infty}^{\infty} R_X(\tau)\epsilon^{-j\omega\tau}\,d\tau \\ &= \mathcal{F}\{R_X(\tau)\} \end{aligned} \tag{6-38}$$

The relationship in (6-38), which is known as the Wiener-Khinchine relation, is of fundamental importance in analyzing random signals because it provides the link between the time domain (correlation function) and the frequency domain (spectral density). Because of the uniqueness of the Fourier transform it follows that the autocorrelation function of a wide-sense stationary random process is the inverse transform of the spectral density. In the case of a nonstationary process the autocorrelation function cannot be recovered from the spectral density—only the time average of the correlation function, as seen from (6-37). In subsequent discussions we will deal only with wide-sense stationary random processes for which (6-38) is valid.

As a simple example of this result, consider an autocorrelation function of the form

$$R_X(\tau) = A\epsilon^{-\beta|\tau|}, \qquad A > 0, \beta > 0 \tag{6-39}$$

The absolute value sign on τ is required by the symmetry of the autocorrelation function. This function is shown in Figure 6-7(a) and is seen to have a discontinuous derivative at $\tau = 0$. Hence, it is necessary to write

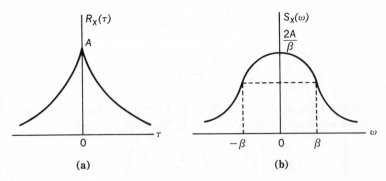

(a)   (b)

Figure 6-7   Relation between (a) autocorrelation function and (b) spectral density.

(6-39) as the sum of two integrals—one for negative values of τ and one for positive values of τ. Thus,

$$
\begin{aligned}
S_X(\omega) &= \int_{-\infty}^{0} A \epsilon^{\beta\tau} \epsilon^{-j\omega\tau} \, d\tau + \int_{0}^{\infty} A \epsilon^{-\beta\tau} \epsilon^{-j\omega\tau} \, d\tau \\
&= A \left. \frac{\epsilon^{(\beta-j\omega)\tau}}{\beta - j\omega} \right|_{-\infty}^{0} + A \left. \frac{\epsilon^{-(\beta+j\omega)\tau}}{-(\beta + j\omega)} \right|_{0}^{\infty} \\
&= A \left[\frac{1}{\beta - j\omega} + \frac{1}{\beta + j\omega} \right] = \frac{2A\beta}{\omega^2 + \beta^2}
\end{aligned}
\tag{6-40}
$$

This spectral density is shown in Figure 6-7(b).

In the stationary case it is also possible to find the autocorrelation function corresponding to a given spectral density by using the inverse Fourier transform. Thus,

$$
R_X(\tau) = \frac{1}{2\pi} \int_{-\infty}^{\infty} S_X(\omega) \epsilon^{j\omega\tau} \, d\omega
\tag{6-41}
$$

An example of the application of this result will be given in the next section.

In obtaining the result in (6-40) the integral was separated into two parts because of the discontinuous slope at the origin. An alternative procedure, which is possible in all cases, is to take advantage of the symmetry of autocorrelation functions. Thus, if (6-38) is written as

$$
S_X(\omega) = \int_{-\infty}^{\infty} R_X(\tau)[\cos \omega\tau - j \sin \omega\tau] \, d\tau
$$

by expressing the exponential in terms of sines and cosines, it may be noted that $R_X(\tau) \sin \omega\tau$ is an odd function of τ and, hence, will integrate to zero. On the other hand, $R_X(\tau) \cos \omega\tau$ is even, and the integral from $-\infty$ to $+\infty$ is just twice the integral from 0 to $+\infty$. Hence,

$$
S_X(\omega) = 2 \int_{0}^{\infty} R_X(\tau) \cos \omega\tau \, d\tau
\tag{6-42}
$$

in an alternative form that does not require integrating over the origin. The corresponding inversion formula, for wide-sense stationary processes, is easily shown to be

$$
R_X(\tau) = \frac{1}{\pi} \int_{0}^{\infty} S_X(\omega) \cos \omega\tau \, d\omega
\tag{6-43}
$$

It was noted earlier that the relationship between spectral density and correlation function could also be expressed in terms of the Laplace transform. However, it should be recalled that the form of the Laplace transform used most often in system analysis requires that the time function being transformed be zero for negative values of time. Autocorrelation functions can never be zero for negative values of τ since they are always even functions of τ. Hence, it is necessary to use the *two-sided*

Laplace transform for this application. The corresponding transform pair may be written as

$$S_X(s) = \int_{-\infty}^{\infty} R_X(\tau)\epsilon^{-s\tau}\, d\tau \qquad \textbf{(6-44)}$$

and

$$R_X(\tau) = \frac{1}{2\pi j} \int_{-j\infty}^{j\infty} S_X(s)\epsilon^{s\tau}\, ds \qquad \textbf{(6-45)}$$

Since the spectral density of a process having a finite mean-square value can have *no* poles on the $j\omega$-axis, the path of integration in (6-45) can always be on the $j\omega$-axis.

Exercise 6-6

A nonstationary random process has an autocorrelation function given by

$$R_X(t, t + \tau) = \sin(4\pi t + 2\pi\tau) + \frac{1}{2}\cos 2\pi\tau$$

Find $S_X(\omega)$.

Answer: $\pi/4[\delta(\omega + 2\pi) + \delta(\omega - 2\pi)]$

6-7   White Noise

The concept of *white noise* was mentioned previously. This term is applied to a spectral density that is constant for all values of ω; that is, $S_X(\omega) = S_0$. It is interesting to determine the correlation function for such a process. This is best done by giving the result and verifying its correctness. Consider an autocorrelation function that is a δ function of the form

$$R_X(\tau) = S_0\delta(\tau)$$

Using this form in (6-38) leads to

$$S_X(\omega) = \int_{-\infty}^{\infty} R_X(\tau)\epsilon^{-j\omega\tau}\, d\tau = \int_{-\infty}^{\infty} S_0\delta(\tau)\epsilon^{-j\omega\tau}\, d\tau = S_0 \qquad \textbf{(6-46)}$$

which is the result for white noise. It is clear, therefore, that the autocorrelation function for white noise is just a δ function with *an area equal to the spectral density.*

It was noted previously that the concept of white noise is fictitious because such a process would have an infinite mean-square value, since the area of the spectral density is infinite. This same conclusion is also apparent from the correlation function. It may be recalled that the mean-square value is equal to the value of the autocorrelation function at $\tau = 0$. For a δ function at the origin, this is also infinite. Nevertheless, the white-noise concept is an extremely valuable one in the analysis of linear systems.   It frequently turns out that the random signal input to a system has a bandwidth that is much greater than the range of frequencies that the system is capable of passing.   Under these circumstances, assuming the input spectral density to be white may greatly simplify the computation of the system response without introducing any significant error.   Examples of this sort will be discussed in Chapters 7 and 8.

Another concept that is frequently used is that of *band-limited white noise*. This implies a spectral density that is constant over a finite bandwidth and zero outside this frequency range.   For example,

$$S_X(\omega) = S_0 \qquad |\omega| \leq 2\pi W$$
$$\qquad\quad = 0 \qquad |\omega| > 2\pi W \tag{6-47}$$

as shown in Figure 6-8(a).   This spectral density is also fictitious even though the mean-square value is finite (in fact, $\overline{X^2} = 2WS_0$).   Why?   It

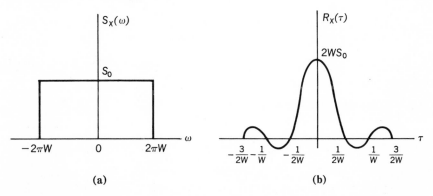

Figure 6-8   Band-limited white noise: (**a**) spectral density and (**b**) autocorrelation function.

can be approached arbitrarily closely, however, and is a convenient form for many analysis problems.

The autocorrelation function for such a process is easily obtained from (6-41).   Thus,

$$R_X(\tau) = \frac{1}{2\pi} \int_{-\infty}^{\infty} S_X(\omega) \epsilon^{j\omega\tau} \, d\omega = \frac{1}{2\pi} \int_{-2\pi W}^{2\pi W} S_0 \epsilon^{j\omega\tau} \, d\omega = \frac{S_0}{2\pi} \frac{\epsilon^{j\omega\tau}}{j\tau} \bigg]_{-2\pi W}^{2\pi W}$$

$$= \frac{S_0}{2\pi} \frac{\epsilon^{j2\pi W\tau} - \epsilon^{-j2\pi W\tau}}{j\tau} = \frac{S_0}{\pi\tau} \sin 2\pi W\tau \tag{6-48}$$

$$= 2WS_0 \left[\frac{\sin 2\pi W\tau}{2\pi W\tau} \right]$$

This is shown in Figure 6-8(b).   Note that in the limit as W approaches infinity this approaches a δ function.

It may be observed from Figure 6-8(b) that random variables from a band-limited process will be uncorrelated if they are separated in time by any multiple of $1/2W$ seconds. It is known also that band-limited functions can be represented exactly and uniquely by a set of samples taken at a rate of twice the bandwidth. This is the so-called *sampling theorem*. Hence, if a band-limited function having a flat spectral density is to be represented by samples, it appears that these samples will be uncorrelated. This lack of correlation among samples may be a significant advantage in carrying out subsequent analysis. In particular, the correlation matrix defined in Section 5-9 for such sampled processes will be a diagonal matrix; that is, all terms not on the major diagonal will be zero.

Exercise 6-7

A spectral density is given by

$$S_X(\omega) = 0.1 \qquad |\omega| \leq 50\pi$$
$$= 0 \qquad |\omega| > 50\pi$$

Find (a) the mean-square value of this process and (b) the smallest value of τ for which the autocorrelation function is zero.

Answer: 0.02, 5

6-8 Cross-spectral Density

When two correlated random processes are being considered, such as the input and the output of a linear system, it is possible to define a pair of quantities known as the cross-spectral densities. For purposes of the present discussion, it is sufficient to simply define them and note a few of their properties without undertaking any formal proofs.

If $F_X(\omega)$ is the Fourier transform of a truncated sample function from one process and $F_Y(\omega)$ is a similar transform from the other process, then the two cross-spectral densities may be defined as

$$S_{XY}(\omega) = \lim_{T \to \infty} \frac{E[F_X(-\omega)F_Y(\omega)]}{2T} \qquad \text{(6-49)}$$

$$S_{YX}(\omega) = \lim_{T \to \infty} \frac{E[F_Y(-\omega)F_X(\omega)]}{2T} \qquad \text{(6-50)}$$

Unlike normal spectral densities, cross-spectral densities need not be real, positive, or even functions of ω. They do have the following properties, however:

1. $S_{XY}(\omega) = S_{YX}{}^*(\omega)$ (* implies complex conjugate)
2. Re $[S_{XY}(\omega)]$ is an even function of ω. Also true for $S_{YX}(\omega)$.
3. Im $[S_{XY}(\omega)]$ is an odd function of ω. Also true for $S_{YX}(\omega)$.

Cross-spectral densities can also be related to cross-correlation functions by the Fourier transform.   Thus for jointly stationary processes,

$$S_{XY}(\omega) = \int_{-\infty}^{\infty} R_{XY}(\tau)\epsilon^{-j\omega\tau}\, d\tau \tag{6-51}$$

$$R_{XY}(\tau) = \frac{1}{2\pi}\int_{-\infty}^{\infty} S_{XY}(\omega)\epsilon^{j\omega\tau}\, d\omega \tag{6-52}$$

$$S_{YX}(\omega) = \int_{-\infty}^{\infty} R_{YX}(\tau)\epsilon^{-j\omega\tau}\, d\tau \tag{6-53}$$

$$R_{YX}(\tau) = \frac{1}{2\pi}\int_{-\infty}^{\infty} S_{YX}(\omega)\epsilon^{j\omega\tau}\, d\omega \tag{6-54}$$

Other properties of cross-spectral densities will be developed as the need arises.

6-9   Measurement of Spectral Density

When random phenomena are encountered in practical situations it is often necessary to measure certain parameters of the phenomena in order to be able to determine how best to design the signal processing system. The case that is most easily handled, and the one that will be considered here, is that in which it can be legitimately assumed that the random process involved is ergodic.   In such cases it is possible to make estimates of various parameters of the process from appropriate time averages. The problems associated with estimating the mean and the correlation function have been considered previously; it is now desired to consider how one may estimate the distribution of power throughout the frequency range occupied by the signal, that is, the spectral density.   This kind of information is invaluable in many engineering applications.   For example, knowing the spectral density of an unwanted or interfering signal often gives valuable clues as to the origin of the signal and may lead to its elimination.   In cases where elimination is not possible, knowledge of the power spectrum often permits design of appropriate filters to reduce the effects of such signals.

As an example of a typical problem of this sort, assume that there is available a continuous recording of a signal $x(t)$ extending over the interval $0 \leq t \leq T$.   The signal $x(t)$ is assumed to be a sample function from an ergodic random process.   It is desired to make an estimate of the spectral density $S_X(\omega)$ of the process from which the recorded signal came.

It might be thought that a reasonable way to find the spectral density would be to find the Fourier transform of the observed sample function and let the square of its magnitude be an estimate of the spectral density. This procedure does not work, however.   Since the Fourier transform of the entire sample function does not even exist, it is not surprising to find that the Fourier transform of a portion of that sample function is a poor estimator of the desired spectral density. This procedure might be possible if one could take an ensemble average of the squared magnitude of the Fourier transform of all (or even some of) the sample functions of the process but since only one sample function is available no such direct approach is possible.

An alternative to the above is to employ the mathematical relationship between the spectral density and the autocorrelation function, as given by (6-38). Since it is possible to estimate autocorrelation functions from a single sample function, as discussed in Section 5-4, the Fourier transform of this estimate will be an estimate of the spectral density. It is this approach that will be discussed here.

It was shown in (5-14) that an estimate of the autocorrelation function of an ergodic process could be obtained from

$$\hat{R}_X(\tau) = \frac{1}{T - \tau} \int_0^{T-\tau} X(t)X(t + \tau)\, dt \qquad 0 \leq \tau \ll T \qquad \text{(5-14)}$$

when $X(t)$ is an arbitrary member of the ensemble. Since τ must be much smaller than T, the length of the record, let the largest permissible value of τ be designated by τ_m. Thus, $\hat{R}_X(\tau)$ has a value given by (5-14) whenever $|\tau| \leq \tau_m$ and is *assumed* to be zero whenever $|\tau| > \tau_m$. A more general way of introducing this limitation on the size of τ is to multiply (5-14) by an even function of τ that is zero when $|\tau| > \tau_m$. Thus, define a new estimate of $R_X(\tau)$ as

$$\begin{aligned}
{}_w\hat{R}_X(\tau) &= \frac{w(\tau)}{T - \tau} \int_0^{T-\tau} X(t)X(t + \tau)\, dt \\
&= w(\tau)\, {}_a\hat{R}_X(\tau)
\end{aligned} \qquad \text{(6-55)}$$

where $w(\tau) = 0$ when $|\tau| > \tau_m$ and is an even function of τ and ${}_a\hat{R}_X(\tau)$ is now *assumed* to exist for all τ. The function $w(\tau)$ is often referred to as a "lag window" since it modifies the estimate of $R_X(\tau)$ by an amount that depends upon the "lag" (that is, the time delay τ) and has a finite width of $2\tau_m$. The purpose of introducing $w(\tau)$, and the choice of a suitable form for it, are extremely important aspects of estimating spectral densities that are all too often overlooked by engineers attempting to make such estimates. The following brief discussion of these topics is hardly adequate to provide a complete understanding, but it may serve to introduce the concepts and to indicate their importance in making meaningful estimates.

Since the spectral density is the Fourier transform of the autocorrelation function, an estimate of spectral density can be obtained by transforming (6-55). Thus,

$$\begin{aligned}
{}_w\hat{S}_X(\omega) &= \mathcal{F}[w(\tau)\, {}_a\hat{R}_X(\tau)] \\
&= \frac{1}{2\pi}\, W(\omega) *\, {}_a\hat{S}_X(\omega)
\end{aligned} \qquad \text{(6-56)}$$

where $W(\omega)$ is the Fourier transform of $w(\tau)$ and the symbol $*$ implies convolution of transforms. $_a\hat{S}_X(\omega)$ is the spectral density associated with $_a\hat{R}_X(\tau)$, which is now defined for all τ but cannot be estimated for all τ.

In order to discuss the purpose of the window function it is important to emphasize that there is a particular window function present even if the problem is ignored. Since (5-14) exists only for $|\tau| \leq \tau_m$, it would be equivalent to (6-55) if a rectangular window defined by

$$w_r(\tau) = 1 \qquad |\tau| \leq \tau_m$$
$$= 0 \qquad |\tau| > \tau_m \qquad \text{(6-57)}$$

were used. Thus, not assigning any window function is really equivalent to using the rectangular window of (6-57). The significance of using a window function of this type can be seen by noting that the corresponding Fourier transform of the rectangular window is

$$\mathfrak{F}[w_r(\tau)] = W_r(\omega) = 2\tau_m \left[\frac{\sin \omega\tau_m}{\omega\tau_m} \right] \qquad \text{(6-58)}$$

and that this transform is negative half the time, as seen in Figure 6-9.

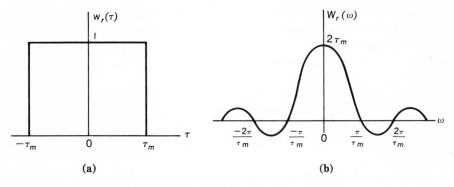

(a)   **(b)**

Figure 6-9 The rectangular-window function and its transform.

Thus, convolving it with $_a\hat{S}_X(\omega)$ can lead to negative values for the estimated spectral density, even though $_a\hat{S}_X(\omega)$ itself is never negative. Thus, the fact that $\hat{R}_X(\tau)$ can be estimated for only a limited range of τ-values (namely, $|\tau| \leq \tau_m$) because of the finite length of the observed data $(\tau_m \ll T)$ may lead to completely erroneous estimates of spectral density, regardless of how accurately $\hat{R}_X(\tau)$ is known within its range of values.

The estimate provided by the rectangular window will be designated

$$_r\hat{S}_X(\omega) = \frac{1}{2\pi} W_r(\omega) *_a\hat{S}_X(\omega) \qquad \text{(6-59)}$$

It should be noted, however, that it is not found by carrying out the convolution indicated (6-59), since $_a\hat{S}_X(\omega)$ cannot be estimated from the limited data available, but instead is just the Fourier transform of $\hat{R}_X(\tau)$ as defined by (5-14).   That is,

$$_r\hat{S}_X(\omega) = \mathfrak{F}[\hat{R}_X(\tau)] \tag{6-60}$$

where

$$\hat{R}_X(\tau) = \frac{1}{T-\tau}\int_0^{T-\tau} X(t)X(t+\tau)\,dt \qquad 0 \leq \tau \leq \tau_m$$

$$= 0 \qquad\qquad\qquad\qquad\qquad \tau > \tau_m$$

and

$$\hat{R}_X(\tau) = \hat{R}_X(-\tau) \qquad\qquad\qquad\qquad \tau < 0$$

Thus, as noted above, $_r\hat{S}_X(\omega)$ is the estimate obtained by ignoring the consequences of the limited range of τ-values.   The problem that now arises is how to modify $_r\hat{S}_X(\omega)$ so as to minimize the erroneous results that occur.   It is this problem that leads to choosing other shapes for the window function $w(\tau)$.

The source of the difficulty associated with $_r\hat{S}_X(\omega)$ is the sidelobes of $W_r(\omega)$.   Clearly, this difficulty could be overcome by selecting a window function that has very small sidelobes in its transform.   One such window function that has been used extensively is the so-called "Hamming window," named after the man who suggested it, and given by

$$w_h(\tau) = 0.54 + 0.46 \cos\frac{\pi\tau}{\tau_m} \qquad |\tau| < \tau_m \tag{6-61}$$

$$= 0 \qquad\qquad\qquad\qquad |\tau| > \tau_m$$

This window and its transform are shown in Figure 6-10.

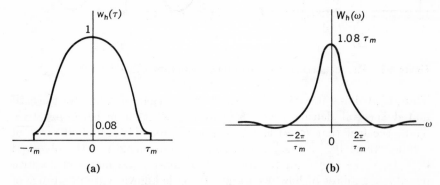

Figure 6-10   (a) The Hamming-window function and (b) its Fourier transform.

The resulting estimate of the spectral density is given formally by

$$_h\hat{S}_X(\omega) = \frac{1}{2\pi}W_h(\omega) *_a\hat{S}_X(\omega) \tag{6-62}$$

but, as before, this convolution cannot be carried out because $_a\hat{S}_X(\omega)$ is not available.   However, if it is noted that

$$w_h(\tau) = \left[0.54 + 0.46 \cos \frac{\pi\tau}{\tau_m} \right] w_r(\tau)$$

then it follows that

$$\mathcal{F}[w_h(t)] = W_h(\omega)$$

$$= \left\{ 0.54\delta(\omega) + 0.23 \left[\delta \left(\omega + \frac{\pi}{\tau_m} \right) + \delta \left(\omega - \frac{\pi}{\tau_m} \right) \right] \right\} * W_r(\omega)$$

since the *nontruncated* constant and cosine term of the Hamming window have Fourier transforms that are δ-functions. Substituting this into (6-62), and utilizing (6-59), leads immediately to

$$_h\hat{S}_X(\omega) = 0.54_r\hat{S}_X(\omega) + 0.23 \left[_r\hat{S}_X \left(\omega + \frac{\pi}{\tau_m} \right) + _r\hat{S}_X \left(\omega - \frac{\pi}{\tau_m} \right) \right] \quad \text{(6-63)}$$

Since $_r\hat{S}_X(\omega)$ can be found, by using (6-60), it follows that (6-63) represents the modification of $_r\hat{S}_X(\omega)$ that is needed to insure that the resulting estimate is always positive.

In the discussion of estimating autocorrelation functions in Section 5-4 it was noted that in almost all practical cases the observed record would be sampled at discrete times of $0, \Delta t, 2\Delta t, \ldots, N\Delta t$ and the resulting estimate formed by the summation:

$$\hat{R}_X(n\Delta t) = \frac{1}{N - n + 1} \sum_{k=0}^{N-n} X_k X_{k+n} \qquad n = 0, 1, \ldots, M \quad \text{(6-64)}$$

Since the autocorrelation function is estimated for discrete values of τ only, it is necessary to perform a discrete approximation to the Fourier transform.   Although there are techniques for doing this that conserve computer time (the so-called "fast Fourier transform"), it is convenient in this discussion to consider the discrete version of (6-41), which is simply the Fourier cosine transform of the autocorrelation function. Thus, the estimated rectangular-window spectral density is

$$_r\hat{S}_X(q\,\Delta\omega) = \Delta t \left[\hat{R}_X(0) + 2 \sum_{n=1}^{M-1} \hat{R}_X(n\,\Delta t) \cos \frac{qn\pi}{M} + \hat{R}_X(M\Delta t) \cos q\pi \right]$$

$$\text{(6-65)}$$

where $q = 0, 1, 2, \ldots, M$

$$\Delta\omega = \frac{\pi}{M\,\Delta t}$$

The corresponding Hamming-window estimate is

$$_h\hat{S}_X(q\,\Delta\omega) = 0.54_r\hat{S}_X(q\,\Delta\omega) + 0.23\{_r\hat{S}_X[(q+1)\,\Delta\omega] + _r\hat{S}_X[(q-1)\,\Delta\omega]\}$$

$$\text{(6-66)}$$

and this represents the final form of the estimate.

Although the problem of evaluating the quality of spectral density estimates is very important, it is also quite difficult. In the first place, Hamming-window estimates are not unbiased, that is, the expected value of the estimate is not the true value of the spectral density. Secondly, it is very difficult to determine the variance of the estimate, although a rough approximation to this variance can be expressed as

$$\text{Var} \left[{}_h\widehat{S}_X(q \, \Delta\omega) \right] \simeq \frac{M}{N} \, S_X{}^2(q\Delta\omega) \qquad \textbf{(6-67)}$$

when $2M\Delta t$ is large enough to include substantially all of the autocorrelation function.

When the spectral density being measured is quite nonuniform over the frequency band, the Hamming-window estimate may give rise to serious errors that can be minimized by "prewhitening," that is, by modifying the spectrum in a known way to make it more nearly uniform. A particularly severe error of this sort arises when the observed data contains a dc component, since this represents a δ-function in the spectral density. In such cases it is very important to remove the dc component from the data before proceeding with the analysis.

Exercise 6-9

It is desired to estimate the spectral density of a random process whose autocorrelation function is

$$\begin{aligned} R_X(\tau) &= 100[1 - 1000|\tau|] & |\tau| &\leq 0.001 \\ &= 0 & |\tau| &> 0.001 \end{aligned}$$

Estimates are to be made at $M = 10$ values of ω and the rms error (standard deviation) of these estimates is to be on the order of 1 percent of the maximum value of the spectral density. How many samples of the time function are needed, what should be the time interval between samples, and what is the frequency range (in hertz) over which the estimates extend?

Answer: 10^{-4}, 10^5, 5000

6-10 Examples and Applications of Spectral Density

The most important application of spectral density is in connection with the analysis of linear systems having random inputs. However, since this application is to be considered in detail in the next chapter, it will not be discussed here. Instead, some examples will be given that emphasize the properties of spectral density and the computational techniques.

The first example will consider the signal in a binary communication system; that is, one in which the message is conveyed by the polarities of a sequence of pulses. The obvious form of pulse to use in such a system is the rectangular one shown in Figure 6-11(a). These pulses all have the same amplitude, but the polarities are either plus or minus with equal probability and are independent from pulse-to-pulse. How-

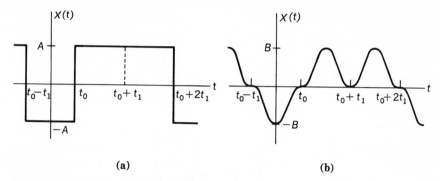

(a)   **(b)**

Figure 6-11 A binary signal with (a) rectangular pulses and (b) raised cosine pulses.

ever, the steep sides of such pulses tend to make this signal occupy more bandwidth than is desirable. An alternative form of pulse is the *raised-cosine* pulse as shown in Figure 6-11(b). The question to be answered concerns the amount by which the bandwidth is reduced by using this type of pulse rather than the rectangular one.

Both of the random processes described above have spectral densities that can be described by the general result in (6-23). In both cases the mean value of pulse amplitude is zero (since each polarity is equally probable), and the variance of pulse amplitude is A^2 for the rectangular pulse and B^2 for the raised-cosine pulse. (See the discussion of delta distributions in Section 2-7.) Thus, all that is necessary is to find $|F(\omega)|^2$ for each pulse shape.

For the rectangular pulse, the shape function $f(t)$ is

$$f(t) = 1 \qquad |t| \leq \frac{t_1}{2}$$

$$= 0 \qquad |t| > \frac{t_1}{2}$$

Hence, its Fourier transform is

$$F(\omega) = \int_{-t_1/2}^{t_1/2} (1)\epsilon^{-j\omega t}\, dt = t_1 \frac{\sin{(\omega t_1/2)}}{(\omega t_1/2)}$$

and, from (6-23), the spectral density of the binary signal is

$$S_X(\omega) = A^2 t_1 \left[\frac{\sin{(\omega t_1/2)}}{\omega t_1/2} \right]^2 \qquad \textbf{(6-68)}$$

which has a maximum value at $\omega = 0$.

For the raised-cosine pulse the shape function is

$$f(t) = \frac{1}{2}\left(1 + \cos\frac{2\pi t}{t_1}\right) \qquad |t| \leq t_1/2$$
$$= 0 \qquad\qquad\qquad |t| > t_1/2$$

The Fourier transform of this shape function becomes

$$F(\omega) = \frac{1}{2}\int_{-t_1/2}^{t_1/2}\left(1 + \cos\frac{2\pi t}{t_1}\right)\epsilon^{-j\omega t}\,dt$$

$$= \frac{t_1}{2}\left[\frac{\sin\,(\omega t_1/2)}{(\omega t_1/2)}\right]\left[\frac{\pi^2}{\pi^2 - (\omega t_1/2)^2}\right]$$

and the corresponding spectral density is

$$S_X(\omega) = \frac{B^2 t_1}{4}\left[\frac{\sin\,(\omega t_1/2)}{\omega t_1/2}\right]^2\left[\frac{\pi^2}{\pi^2 - (\omega t_1/2)^2}\right]^2 \qquad\text{(6-69)}$$

which has a maximum value at $\omega = 0$.

In evaluating the bandwidths of these spectral densities there are many different criteria that might be employed. However, when one wishes to reduce the interference between two communication systems it is reasonable to consider the bandwidth outside of which the signal spectral density is below some specified fraction (say 1 percent) of the maximum spectral density. That is, one wishes to find the value of ω_1 such that

$$\frac{S_X(\omega)}{S_X(0)} \leq 0.01 \qquad |\omega| > \omega_1$$

Since $\sin\,(\omega t_1/2)$ can never be larger than 1, this condition will be assured for (6-68) when

$$\frac{A^2 t_1\left[\dfrac{1}{(\omega t_1/2)}\right]^2}{A^2 t_1} \leq 0.01$$

from which

$$\omega_1 \simeq \frac{20}{t_1}$$

for the case of rectangular pulses. When raised-cosine pulses are used, this condition becomes

$$\frac{B^2 t_1/4[1/(\omega t_1/2)]^2[\pi^2/(\pi^2 - (\omega t_1/2)^2)]^2}{B^2 t_1/4} \leq 0.01$$

This leads to

$$\omega_1 \simeq \frac{10.68}{t_1}$$

It is clear that the use of raised-cosine pulses, rather than rectangular pulses, has cut the bandwidth almost in half (when bandwidth is specified according to this criterion).

Almost all of the examples of spectral density that have been considered throughout this chapter have been low-pass functions; that is, the spectral density has had its maximum value at $\omega = 0$. However, many practical situations arise in which the maximum value of spectral density occurs at some high frequency, and the second example will illustrate a situation of this sort. Figure 6-12 shows a typical band-pass

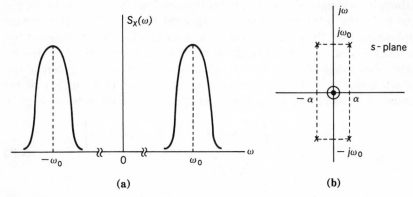

(a)   **(b)**

Figure 6-12   (a) A band-pass spectral density and (b) the corresponding pole-zero plot.

spectral density and the corresponding pole-zero configuration. The complex frequency representation for this spectral density is obtained easily from the pole-zero plot. Thus,

$$S_X(s) = \frac{S_0(s)(-s)}{(s + \alpha + j\omega_0)(s + \alpha - j\omega_0)(s - \alpha + j\omega_0)(s - \alpha - j\omega_0)}$$

$$= \frac{-S_0 s^2}{[(s + \alpha)^2 + \omega_0^2][(s - \alpha)^2 + \omega_0^2]} \qquad \text{(6-70)}$$

where S_0 is a scale factor. Note that this spectral density is zero at zero frequency.

The mean-square value associated with this spectral density can be obtained by either of the methods discussed in Section 6-5. If Table 6-1 is used, it is seen readily that

$$c(s) = s \qquad c_1 = 1 \qquad c_0 = 0$$
$$d(s) = s^2 + 2\alpha s + \alpha^2 + \omega_0^2 \qquad d_2 = 1, d_1 = 2\alpha, d_0 = \alpha^2 + \omega_0^2$$

The mean-square value is then related to I_2 by

$$\overline{X^2} = S_0 I_2 = S_0 \frac{c_1^2 d_0 + c_0^2 d_2}{2 d_0 d_1 d_2} = S_0 \frac{(1)^2(\alpha^2 + \omega_0^2) + 0}{2(\alpha^2 + \omega_0^2)(2\alpha)(1)}$$

$$= \frac{S_0}{4\alpha}$$

An interesting result of this calculation is that the mean-square value depends only upon the bandwidth parameter α and not upon the center frequency ω_0.

The third example concerns the physical interpretation of spectral density as implied by the relation

$$\overline{X^2} = \frac{1}{2\pi} \int_{-\infty}^{\infty} S_X(\omega)\, d\omega$$

Although this expression only relates the total mean-square value of the process to the total area under the spectral density, there is a further implication that the mean-square value associated with any range of frequencies is similarly related to the partial area under the spectral density within that range of frequencies. That is, if one chooses any pair of frequencies, say ω_1 and ω_2, the mean-square value of that portion of the random process having energy between these two frequencies is

$$\overline{{}_1X_2{}^2} = \frac{1}{2\pi} \left[\int_{-\omega_2}^{-\omega_1} S_X(\omega)\, d\omega + \int_{\omega_1}^{\omega_2} S_X(\omega)\, d\omega \right]$$

$$= \frac{1}{\pi} \int_{\omega_1}^{\omega_2} S_X(\omega)\, d\omega \tag{6-71}$$

The second form in (6-71) is a consequence of $S_X(\omega)$ being an even function of ω.

As an illustration of this concept, consider again the spectral density derived in (6-40). This was

$$S_X(\omega) = \frac{2A\beta}{\omega^2 + \beta^2} \tag{6-40}$$

where A is the total mean-square value of the process. Suppose it is desired to find the frequency above which one-half the total mean-square value (or average power) exists. This means that we want to find the ω_1 (with $\omega_2 = \infty$) for which

$$\frac{1}{\pi} \int_{\omega_1}^{\infty} \frac{2A\beta}{\omega^2 + \beta^2}\, d\omega = \frac{1}{2} \left[\frac{1}{\pi} \int_{0}^{\infty} \frac{2A\beta}{\omega^2 + \beta^2}\, d\omega \right] = \frac{1}{2} A$$

Thus,

$$\int_{\omega_1}^{\infty} \frac{d\omega}{\omega^2 + \beta^2} = \frac{\pi}{4\beta}$$

since the A cancels out. The integral becomes

$$\frac{1}{\beta} \tan^{-1} \frac{\omega}{\beta} \Big|_{\omega_1}^{\infty} = \frac{1}{\beta} \left[\frac{\pi}{2} - \tan^{-1} \frac{\omega_1}{\beta} \right] = \frac{\pi}{4\beta}$$

from which

$$\tan^{-1} \frac{\omega_1}{\beta} = \frac{\pi}{4}$$

and

$$\omega_1 = \beta$$

Thus, one-half of the average power of this process occurs at frequencies above β and one-half below β. Note that in this particular case, β is also the frequency at which the spectral density is one-half of its maximum value at $\omega = 0$. This result is peculiar to this particular spectral density and is not true in general. For example, in the band-limited white-noise case shown in Figure 6-8, the spectral density reaches one-half of its maximum value at $\omega = 2\pi W$, but one-half of the average power occurs at frequencies greater than $\omega = \pi W$. These conclusions are obvious from the sketch.

Exercise 6-10

For the spectral density of (6-40), find the value of ω_1 such that one-tenth of the average power lies outside of ω_1.

Answer: 6.30β

■ PROBLEMS

6-1 For each of the following functions of ω, state whether it can or cannot be a valid expression for the spectral density of a random process. If it cannot, state why.

a. $\dfrac{\omega^2 + 9}{\omega^4 - 3\omega^2 + 2}$

d. $\dfrac{\sqrt{\omega^4 - 2}}{\omega^6 + 10\omega^4 + 2\omega^2 + 1}$

b. $\dfrac{\omega^2 + 10}{\omega^4 + 5\omega^2 + 4}$

e. $\left(\dfrac{\sin \omega}{\omega}\right)^2$

c. $\dfrac{1}{\omega^2 + 2\omega + 1}$

f. $\delta(\omega) + \dfrac{\omega^2}{\omega^4 + 1}$

6-2 Using the result given in Equation (6-21) write an expression for the spectral density of a random process whose sample functions are given by the Fourier series expansion

$$X(t) = \frac{a_0}{2} + \sum_{n=1}^{\infty} a_n \cos n\omega_0(t + t_0) + b_n \sin n\omega_0(t + t_0)$$

where t_0 is uniformly distributed over one period and a_n and b_n are constants.

6-3 A random process has periodic sample functions as shown where A

is a constant and t_0 is a random variable uniformly distributed between O and T.   Using Equation (6-23)

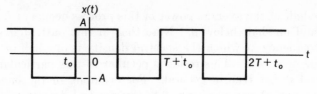

a. Find the sketch and spectral density $S_X(\omega)$.
b. Repeat for the process whose sample functions are $Y(t) = A + X(t)$.

6-4   A polarity sampler is shown in the accompanying figure.   Assume that the sampler makes an instantaneous determination of the polarity of the input and then generates a positive or negative unit amplitude pulse at the output depending on whether the sample polarity was positive or negative.   If a zero mean random noise whose bandwidth is large

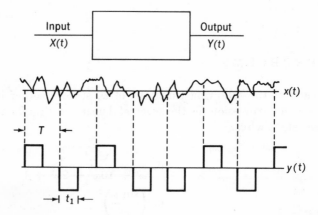

compared to the sampling frequency is applied to the polarity sampler each sample will be independent of all others and the output will be a random amplitude binary waveform.   Find the spectral density of the output in terms of the sampling period, T, and the duty factor, t_1/T.

6-5   For each of the acceptable spectral densities in Problem 6-1, write the corresponding complex frequency representation $S_X(s)$.

6-6   Compute the mean square values for each of the acceptable spectral densities given in Problem 6-1.

6-7   Find the mean-square value of the stationary random process whose spectral density is

$$S_X(s) = \frac{s^2}{s^6 - 1}$$

6-8   A stationary random process has an autocorrelation function given by

$$R_X(\tau) = A e^{-\alpha|\tau|} \cos \omega_0 \tau$$

Find and sketch the spectral density of the random process.

6-9   A random process has sample functions as shown where t_0 is a random variable uniformly distributed between 0 and T and the a_k are independent random variables uniformly distributed between -1 and $+1$.

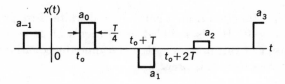

Find the spectral density of this process by first finding the autocorrelation function and then computing the Fourier transform.   Check the result using Equation (6-23).

6-10   Consider two stationary random processes $X(t)$ and $Y(t)$.   Write $S_{XY}(\omega)$ and $S_{YX}(\omega)$ in terms of their real and imaginary parts and show that

$$\text{Re } S_{XY}(\omega) = \text{Re } S_{YX}(\omega)$$
$$\text{Im } S_{XY}(\omega) = - \text{Im } S_{YX}(\omega)$$

6-11   A new random process, $Z(t)$, is formed from the sum of two jointly stationary random processes, $X(t)$ and $Y(t)$.

a. Compute the spectral density of the new process $S_Z(\omega)$.
b. If $X(t)$ and $Y(t)$ are statistically independent with means of $\bar{X}$ and $\bar{Y}$, what can be said of $S_{XY}(\omega)$ and $S_{YX}(\omega)$?
c. Compute the cross spectral density $S_{XZ}(\omega)$.

6-12   One of the earliest window functions employed in smoothing of spectra was the so-called "Hanning window" defined as

$$w(\tau) = 0.5 + 0.5 \cos \frac{\pi\tau}{\tau_m} \qquad |\tau| < \tau_m$$
$$= 0 \qquad |\tau| > \tau_m$$

The reason for using this window is the very simple relationship that results when it is applied to the rectangular window spectrum obtained with no smoothing.

a. Derive an expression similar to Equation (6-63) for the Hanning window.
b. Compare the sidelobe levels of the Hamming and Hanning windows.

6-13 Compare the performance of a triangular pulse with that of the rectangular and raised cosine pulses for use in a binary communication system (see Section 6-10).   Assume that the energy in all pulses is the same.

6-14 For a rectangular pulse shape such as might be used for binary communication (see Problem 6-13) compute the frequency below which 90 percent of the power is contained.   Repeat for 99 percent of the power. (Hint: Use numerical or graphical integration.)

6-15 A sample function from a band-limited white noise process, having a bandwidth of W Hz, is sampled at intervals of $\Delta t = 1/2W$.   If the spectral density of this process is S_0, determine the covariance matrix for N samples taken from this process.

■ REFERENCES

See References for Chapter 1, particularly Beckmann, Davenport and Root, Papoulis, and Thomas.   The following additional reference provides considerable elaboration on the techniques of estimating power spectra.

Blackman, R. B. and J. W. Tukey, *The Measurement of Power Spectra.* New York: Dover Publications, 1958.

Response of Linear Systems to Random Inputs

7-1 Introduction

The discussion in the preceding chapters has been devoted to finding suitable mathematical representations for random functions of time. The next step is to see how these mathematical representations can be used to determine the response, or output, of a linear system when the input is a random signal rather than a deterministic one.

It is assumed that the student is already familiar with the usual methods of analyzing linear systems in either the time domain or the frequency domain. These methods will be restated here in order to clarify the notation, but no attempt will be made to review all of the essential concepts. The system itself will be represented either in terms of its *impulse response* $h(t)$ or its *system function* $H(\omega)$, which is just the Fourier transform of the impulse response. It will be convenient in many cases also to use the *transfer function* $H(s)$, which is the Laplace transform of the impulse response. In most cases the initial conditions will be assumed to be zero, for convenience, but any nonzero initial conditions can be taken into account by the usual methods if necessary.

When the input to a linear system is deterministic either approach will lead to a unique relationship between the input and output. When the input to the system is a sample function from a random process there is again a unique relationship between the excitation and the response; however, because of its random nature, we do not have an explicit representation of the excitation and, therefore, cannot obtain an explicit expression for the response. In this case we must be content with either a probabilistic or a statistical description of the response, just as we must use this type of description for the random excitation itself.[1] Of these two approaches, statistical and probabilistic, the statistical approach is the most useful. In only a very limited class of problems is it possible to obtain a probabilistic description of the output based on a probabilistic description of the input, whereas in many cases of interest a statistical model of the output can be obtained readily by performing simple mathematical operations on the statistical model of the input. With the statistical method, such quantities as the mean, correlation function, and spectral density of the output can be determined. Only the statistical approach will be considered in the following sections.

7-2   Analysis in the Time Domain

By means of the convolution integral it is possible to determine the response of a linear system to a very general excitation. In the case of time-varying systems or nonstationary random excitations, or both, the details become quite involved; therefore, these cases will not be considered here. To make the analysis more realistic we will further restrict our considerations to physically realizable systems that are bounded-input/bounded-output stable. If the input time function is designated

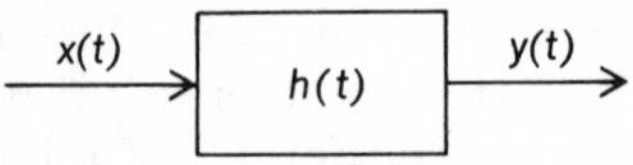

Figure 7-1   Time-domain representation of a linear system.

as $x(t)$, the system impulse response as $h(t)$, and the output time function as $y(t)$, as shown in Figure 7-1, then they are all related either by

$$y(t) = \int_0^\infty x(t - \lambda)h(\lambda)\, d\lambda \tag{7-1}$$

or by

$$y(t) = \int_{-\infty}^t x(\lambda)h(t - \lambda)\, d\lambda \tag{7-2}$$

---

[1] By a probabilistic description we mean one in which certain probability functions are specified; by a statistical description we mean one in which certain ensemble averages are specified (for example, mean, variance, correlation).

The physical realizability and stability constraints on the system are given by

$$h(t) = 0 \qquad t < 0 \tag{7-3}$$

$$\int_{-\infty}^{\infty} |h(t)| \, dt < \infty \tag{7-4}$$

Starting from these specifications, many important characteristics of the output of a system excited by a stationary random process can be determined.

7-3   Mean and Mean-Square Value of System Output

The most convenient form of the convolution integral, when the input $X(t)$ is a sample function from a random process, is

$$Y(t) = \int_{0}^{\infty} X(t - \lambda)h(\lambda) \, d\lambda \tag{7-5}$$

since the limits of integration do not depend on t. Using this form, consider first the mean value of the output $y(t)$. This is given by

$$\bar{Y} = E[Y(t)] = E\left[\int_{0}^{\infty} X(t - \lambda)h(\lambda) \, d\lambda \right] \tag{7-6}$$

The next logical step is to interchange the sequence in which the time integration and the expectation are performed; that is, to move the expectation operation inside the integral. Before doing this, however, it is necessary to digress a moment and consider the conditions under which such an interchange is justified.

The problem of finding the expected value of an integral whose integrand contains a random variable arises many times. In almost all such cases it is desirable to be able to move the expectation operation inside the integral and thus simplify the integrand. Fortunately, this interchange is possible in almost all cases of practical interest and, hence, is used throughout this book with little or no comment. It is advisable, however, to be aware of the conditions under which this is possible, even though the reasons for these conditions are not fully understood. The conditions may be stated as follows:

If $Z(t)$ is a sample function from a random process (or some function, such as the square, of the sample function) and $f(t)$ is a nonrandom time function, then

$$E\left[\int_{t_1}^{t_2} Z(t)f(t) \, dt \right] = \int_{t_1}^{t_2} E[Z(t)]f(t) \, dt$$

if

1. $\int_{t_1}^{t_2} E[|Z(t)|]|f(t)| \, dt < \infty$

and

>   **2.** $Z(t)$ is bounded on the interval t_1 to t_2. Note that t_1 and t_2 may be infinite. (There is *no* requirement that $Z(t)$ be from a stationary process.)

In applying this result to the analysis of linear systems the nonrandom function $f(t)$ is usually the impulse response $h(t)$. For wide-sense stationary input random processes the quantity $E[|Z(t)|]$ is a *constant* not dependent on time t. Hence, the stability condition of (7-4) is sufficient to satisfy condition 1. The boundedness of $Z(t)$ is always satisfied by physical signals, although there are some mathematical representations that may not be bounded.

Returning to the problem of finding the mean value of the output of a linear system, it follows that

$$\bar{Y} = \int_0^\infty E[X(t - \lambda)]h(\lambda)\, d\lambda = \bar{X} \int_0^\infty h(\lambda)\, d\lambda \qquad \text{(7-7)}$$

when the input process is wide-sense stationary. It should be recalled from earlier work in systems analysis that the *area* of the impulse response is just the *dc gain* of the system—that is, the transfer function of the system evaluated at $\omega = 0$. Hence, (7-7) simply states the obvious fact that the dc component of the output is equal to the dc component of the input times the dc gain of the system. If the input random process has zero mean, the output process will also have zero mean. If the system does not pass direct current, the output process will always have zero mean.

In order to find the mean-square value of the output we must be able to calculate the mean value of the *product* of two integrals. However, such a product may always be written as an iterated double integral if the variables of integration are kept distinct. Thus,

$$
\begin{aligned}
\overline{Y^2} = E[Y^2(t)] &= E\left[\int_0^\infty X(t - \lambda_1)h(\lambda_1)\, d\lambda_1 \cdot \int_0^\infty X(t - \lambda_2)h(\lambda_2)\, d\lambda_2 \right] \\
&= E\left[\int_0^\infty d\lambda_1 \int_0^\infty X(t - \lambda_1)X(t - \lambda_2)h(\lambda_1)h(\lambda_2)\, d\lambda_2 \right] \qquad \text{(7-8)} \\
&= \int_0^\infty d\lambda_1 \int_0^\infty E[X(t - \lambda_1)X(t - \lambda_2)]h(\lambda_1)h(\lambda_2)\, d\lambda_2 \qquad \text{(7-9)}
\end{aligned}
$$

in which the subscripts on λ_1 and λ_2 have been introduced to keep the variables of integration distinct. The expected value inside the double integral is simply the autocorrelation function for the input random process; that is,

$$E[X(t - \lambda_1)X(t - \lambda_2)] = R_X(t - \lambda_1 - t + \lambda_2) = R_X(\lambda_2 - \lambda_1)$$

Hence, (7-9) becomes

$$\overline{Y^2} = \int_0^\infty d\lambda_1 \int_0^\infty R_X(\lambda_2 - \lambda_1)h(\lambda_1)h(\lambda_2)\, d\lambda_2 \qquad \text{(7-10)}$$

Although (7-10) is usually not difficult to evaluate, since both $R_X(\tau)$ and $h(t)$ are likely to contain only exponentials, it is frequently very tedious to carry out the details.   This is because such autocorrelation functions often have discontinuous derivatives at the origin (which in this case is $\lambda_1 = \lambda_2$) and thus the integral must be broken up into several ranges.   This point will be illustrated later.   At the moment, however, it is instructive to consider a much simpler situation—one in which the input is a sample function from white noise.   For this case, it was shown in Section 6-7 that

$$R_X(\tau) = S_0\delta(\tau)$$

where S_0 is the spectral density of the white noise.   Hence (7-10) becomes

$$\overline{Y^2} = \int_0^\infty d\lambda_1 \int_0^\infty S_0\delta(\lambda_2 - \lambda_1)h(\lambda_1)h(\lambda_2)\, d\lambda_2 \qquad \textbf{(7-11)}$$

Integrating over λ_2 yields

$$\overline{Y^2} = S_0 \int_0^\infty h^2(\lambda)\, d\lambda \qquad \textbf{(7-12)}$$

Hence, for this case it is the area of the *square* of the impulse response that is significant.[2]

As a means of illustrating some of these ideas with a simple example, consider the single-section, low-pass *RC* circuit shown in Figure 7-2.

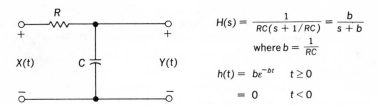

Figure 7-2   Simple RC circuit and its impulse response.

The mean value of the output is, from (7-7),

$$\bar{Y} = \bar{X} \int_0^\infty b\epsilon^{-b\lambda}\, d\lambda = \bar{X}b \left.\frac{\epsilon^{-b\lambda}}{-b}\right|_0^\infty = \bar{X} \qquad \textbf{(7-13)}$$

This result is obviously correct, since it is apparent by inspection that the dc gain of this circuit is unity.

Next consider the mean-square value of the output when the input is

---

[2] It should be noted that, for some functions, this integral can diverge even when (7-4) is satisfied.   This occurs, for instance, whenever $h(t)$ contains δ functions.   The high-pass *RC* circuit is an example of this.

white noise.   From (7-12), this is

$$\overline{Y^2} = S_0 \int_0^\infty b^2 \epsilon^{-2b\lambda} \, d\lambda = b^2 S_0 \left. \frac{\epsilon^{-2b\lambda}}{-2b} \right|_0^\infty = \frac{bS_0}{2} \qquad \textbf{(7-14)}$$

Note that the parameter b, which is the reciprocal of the time constant, is also related to the half-power bandwidth of the system. In particular, this bandwidth B is

$$B = \frac{1}{2\pi RC} = \frac{b}{2\pi} \text{ Hz}$$

so that (7-14) could be written as

$$\overline{Y^2} = \pi B S_0 \qquad \textbf{(7-15)}$$

It is evident from this that the mean-square value of the output of this system increases *linearly* with the bandwidth of the system. This is a typical result whenever the bandwidth of the input random process is large compared with the bandwidth of the system.

Exercise 7-3

A system has an impulse response of

$$\begin{aligned} h(t) &= 2\epsilon^{-t} - \epsilon^{-2t} & t \geq 0 \\ &= 0 & t < 0 \end{aligned}$$

If white noise having a spectral density of $12 \text{ V}^2/\text{Hz}$ is applied to the input of this system, find the (a) mean and (b) mean-square values of the output.

Answer: 0, 11

7-4   Autocorrelation Function of System Output

A problem closely related to that of finding the mean-square value is the determination of the autocorrelation function at the output of the system. By definition, this autocorrelation function is

$$R_Y(\tau) = E[Y(t)Y(t + \tau)]$$

Following the same steps as in (7-9), except for replacing t by $t + \tau$ in one factor, the autocorrelation function may be written as

$$R_Y(\tau) = \int_0^\infty d\lambda_1 \int_0^\infty E[X(t - \lambda_1)X(t + \tau - \lambda_2)]h(\lambda_1)h(\lambda_2) \, d\lambda_2 \quad \textbf{(7-16)}$$

In this case the expected value inside the integral is

$$\begin{aligned} E[X(t - \lambda_1)X(t + \tau - \lambda_2)] \\ = R_X(t - \lambda_1 - t - \tau + \lambda_2) = R_X(\lambda_2 - \lambda_1 - \tau) \end{aligned}$$

Hence, the output autocorrelation function becomes

$$R_Y(\tau) = \int_0^\infty d\lambda_1 \int_0^\infty R_X(\lambda_2 - \lambda_1 - \tau) h(\lambda_1) h(\lambda_2) \, d\lambda_2 \qquad \text{(7-17)}$$

Note the similarity between this result and that for the mean-square value. In particular, for $\tau = 0$, this reduces exactly to (7-10), as it must.

For the special case of white noise into the system, the expression for the output autocorrelation function becomes much simpler. Let

$$R_X(\tau) = S_0 \delta(\tau)$$

as before, and substitute into (7-17). Thus,

$$R_Y(\tau) = \int_0^\infty d\lambda_1 \int_0^\infty S_0 \delta(\lambda_2 - \lambda_1 - \tau) h(\lambda_1) h(\lambda_2) \, d\lambda_2$$

$$= S_0 \int_0^\infty h(\lambda_1) h(\lambda_1 + \tau) \, d\lambda_1 \qquad \text{(7-18)}$$

Hence, for the white-noise case, the output autocorrelation function is proportional to the time correlation function of the impulse response.

This point can be illustrated by means of the linear system of Figure 7-2 and a white-noise input. Thus,

$$R_Y(\tau) = S_0 \int_0^\infty (b\epsilon^{-b\lambda}) b\epsilon^{-b(\lambda+\tau)} \, d\lambda$$

$$= b^2 S_0 \epsilon^{-b\tau} \left. \frac{\epsilon^{-2b\lambda}}{-2b} \right|_0^\infty = \frac{bS_0}{2} \epsilon^{-b\tau} \qquad \tau \geq 0 \qquad \text{(7-19)}$$

This result is valid only for $\tau \geq 0$. When $\tau < 0$, the range of integration must be altered because the impulse response is always zero for negative values of the argument. The situation can be made clearer by means of the two sketches shown in Figure 7-3, which show the factors in the inte-

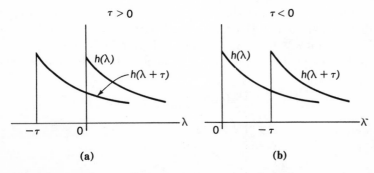

Figure 7-3 Factors in the integrand of (7-18) when the RC circuit of Figure 7-2 is used.

grand of (7-18) for both ranges of τ. The integrand is zero, of course, when either factor is zero.   When $\tau < 0$, the integral becomes

$$R_Y(\tau) = S_0 \int_{-\tau}^{\infty} (b\epsilon^{-b\lambda})b\epsilon^{-b(\lambda+\tau)} \, d\lambda$$

$$= b^2 S_0 \epsilon^{-b\tau} \frac{\epsilon^{-2b\lambda}}{-2b} \Big|_{-\tau}^{\infty} = \frac{bS_0}{2} \epsilon^{b\tau} \qquad \tau \leq 0 \qquad \textbf{(7-20)}$$

From (7-19) and (7-20), the complete autocorrelation function can be written as

$$R_Y(\tau) = \frac{bS_0}{2} \epsilon^{-b|\tau|} \qquad -\infty < \tau < \infty \qquad \textbf{(7-21)}$$

It is now apparent that the calculation for $\tau < 0$ was needless.   Since the autocorrelation function is an even function of τ, the complete form could have been obtained immediately from the case for $\tau \geq 0$.   This procedure will be followed in the future.

It is desirable to consider at least one example in which the input random process is not white.   In so doing, it will be possible to illustrate some of the integration problems that develop and, at the same time, use the results to infer something about the validity and usefulness of the white-noise approximation.   For this purpose, assume that the input random process to the RC circuit of Figure 7-2 has an autocorrelation function of the form

$$R_X(\tau) = \frac{\beta S_0}{2} \epsilon^{-\beta|\tau|} \qquad -\infty < \tau < \infty \qquad \textbf{(7-22)}$$

The coefficient $\beta S_0/2$ has been selected so that this random process has a spectral density at $\omega = 0$ of S_0; see (6-40) and Figure 6-7(b).   Thus, at low frequencies the spectral density is the same as the white-noise spectrum previously assumed.

In order to determine the appropriate ranges of integration, it is desirable to look at the autocorrelation function $R_X(\lambda_2 - \lambda_1 - \tau)$, as a function of λ_2 for $\tau > 0$.   This is shown in Figure 7-4.   Since λ_2 is always positive for the evaluation of (7-17), it is clear that the ranges of inte-

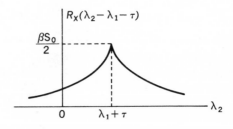

Figure 7-4   Autocorrelation function to be used in (7-17).

gration should be from 0 to $(\lambda_1 + \tau)$ and from $(\lambda_1 + \tau)$ to ∞.   Hence, (7-17) may be written as

$$
R_Y(\tau) = \int_0^\infty d\lambda_1 \int_0^{\lambda_1+\tau} R_X(\lambda_2 - \lambda_1 - \tau)h(\lambda_1)h(\lambda_2)\, d\lambda_2
$$

$$
+ \int_0^\infty d\lambda_1 \int_{\lambda_1+\tau}^\infty R_X(\lambda_2 - \lambda_1 - \tau)h(\lambda_1)h(\lambda_2)\, d\lambda_2
$$

$$
= \frac{b^2\beta S_0}{2} \int_0^\infty \epsilon^{-(b+\beta)\lambda_1}\, d\lambda_1 \int_0^{\lambda_1+\tau} \epsilon^{-\beta\tau}\epsilon^{-(b-\beta)\lambda_2}\, d\lambda_2
$$

$$
+ \frac{b^2\beta S_0}{2} \int_0^\infty \epsilon^{-(b-\beta)\lambda_1}\, d\lambda_1 \int_{\lambda_1+\tau}^\infty \epsilon^{\beta\tau}\epsilon^{-(b+\beta)\lambda_2}\, d\lambda_2 \qquad \textbf{(7-23)}
$$

$$
= \frac{b^2\beta S_0}{-2(b-\beta)}\, \epsilon^{-\beta\tau} \int_0^\infty \epsilon^{-(b+\beta)\lambda_1}[\epsilon^{-(b-\beta)(\lambda_1+\tau)} - 1]\, d\lambda_1
$$

$$
- \frac{b^2\beta S_0}{-2(b+\beta)}\, \epsilon^{\beta\tau} \int_0^\infty \epsilon^{-(b-\beta)\lambda_1}\epsilon^{-(b+\beta)(\lambda_1+\tau)}\, d\lambda_1
$$

$$
= \frac{b^2\beta S_0}{2(b-\beta)}\left(-\frac{\epsilon^{-b\tau}}{2b} + \frac{\epsilon^{-\beta\tau}}{b+\beta}\right) + \frac{b^2\beta S_0}{2(b+\beta)}\left(\frac{\epsilon^{-b\tau}}{2b}\right)
$$

$$
= \frac{b^2\beta S_0}{2(b^2 - \beta^2)}\left(\epsilon^{-\beta\tau} - \frac{\beta}{b}\epsilon^{-b\tau}\right) \qquad \tau > 0
$$

From symmetry, the expression for $\tau < 0$ can be written directly.   The final result is

$$
R_Y(\tau) = \frac{b^2\beta S_0}{2(b^2 - \beta^2)}\left(\epsilon^{-\beta|\tau|} - \frac{\beta}{b}\epsilon^{-b|\tau|}\right) \qquad \textbf{(7-24)}
$$

In order to compare this result with the previously obtained result for white noise at the input, it is only necessary to let β approach infinity. In this case,

$$
\lim_{\beta \to \infty} R_Y(\tau) = \frac{bS_0}{2}\, \epsilon^{-b|\tau|} \qquad \textbf{(7-25)}
$$

which is exactly the same as (7-21).   Of greater interest, however, is the case when β is large compared to b but still finite.   This corresponds to the physical situation in which the bandwidth of the input random process is large compared to the bandwidth of the system.   In order to make this comparison, write (7-24) as

$$
R_Y(\tau) = \frac{bS_0}{2}\, \epsilon^{-b|\tau|}\left[\frac{1}{1 - (b^2)/\beta^2}\right]\left[1 - \frac{b}{\beta}\epsilon^{-(\beta-b)|\tau|}\right] \qquad \textbf{(7-26)}
$$

The first factor in (7-26) is the autocorrelation function of the output when the input is white noise. The second factor is the one by which the true autocorrelation of the system output differs from the white-noise approximation. It is clear that as β becomes large compared to b, this factor approaches unity.

The point to this discussion is that there are many practical situations in which the input noise has a bandwidth that is much greater than the system bandwidth, and in these cases it is quite reasonable to use the white-noise approximation. In doing so, there is a great saving in labor without much loss in accuracy; for example, in a high-gain vacuum-tube amplifier with a bandwidth of 10 MHz, the most important source of noise is shot noise in the first stage, which may have a bandwidth of 1000 MHz. Hence, the factor b/β in (7-26), assuming that this form applies, will be only 0.01, and the error in using the white-noise approximation will not exceed 1 percent.

Exercise 7-4

White noise having a spectral density of 6 V²/Hz is at the input to a system having an impulse response of

$$h(t) = 1 - t \qquad 0 \le t \le 1$$
$$= 0 \qquad \text{elsewhere}$$

Find the value of the autocorrelation function at the system output at (a) $\tau = 0$, (b) $\frac{1}{2}$, and (c) 1.

Answer: $0, \frac{5}{8}, 2$

7-5 Cross-correlation Between Input and Output

When a sample function from a random process is applied to the input of a linear system, the output must be related in some way to the input. Hence, they will be correlated, and the nature of the cross-correlation function is important. In fact, it will be shown very shortly that this relationship can be used to provide a practical technique for measuring the impulse response of any linear system.

One of the cross-correlation functions for input and output is defined by

$$R_{XY}(\tau) = E[X(t)Y(t + \tau)] \qquad \text{(7-27)}$$

which, in integral form, becomes

$$R_{XY}(\tau) = E\left[X(t)\int_0^\infty X(t + \tau - \lambda)h(\lambda)\,d\lambda\right] \qquad \text{(7-28)}$$

Since $X(t)$ is not a function of λ, it may be moved inside the integral and then the expectation may be moved inside. Thus,

$$R_{XY}(\tau) = \int_0^\infty E[X(t)X(t+\tau-\lambda)]h(\lambda)\,d\lambda$$

$$= \int_0^\infty R_X(\tau-\lambda)h(\lambda)\,d\lambda$$

(7-29)

Hence, this cross-correlation function is just the convolution of the input autocorrelation function and the impulse response of the system.

The other cross-correlation function is

$$R_{YX}(\tau) = E[X(t+\tau)Y(t)] = E\left[X(t+\tau)\int_0^\infty X(t-\lambda)h(\lambda)\,d\lambda\right]$$

$$= \int_0^\infty E[X(t+\tau)X(t-\lambda)]h(\lambda)\,d\lambda$$

$$= \int_0^\infty R_X(\tau+\lambda)h(\lambda)\,d\lambda$$

(7-30)

Since the autocorrelation function in (7-30) is symmetrical about $\lambda = -\tau$ and the impulse response is zero for negative values of λ, this cross-correlation function will *always* be different from $R_{XY}(\tau)$. They will, however, have the same value at $\tau = 0$.

The above results become even simpler when the input to the system is considered to be a sample function of white noise. For this case,

$$R_X(\tau) = S_0\delta(\tau)$$

and

$$R_{XY}(\tau) = \int_0^\infty S_0\delta(\tau-\lambda)h(\lambda)\,d\lambda = S_0h(\tau) \qquad \tau \geq 0$$

$$= 0 \qquad \tau < 0$$

(7-31)

Likewise,

$$R_{YX}(\tau) = \int_0^\infty S_0\delta(\tau+\lambda)h(\lambda)\,d\lambda = 0 \qquad \tau > 0$$

$$= S_0h(-\tau) \qquad \tau \leq 0$$

(7-32)

It is the result shown in (7-31) that leads to the procedure for measuring the impulse response, which will be discussed next.

Consider the block diagram shown in Figure 7-5. The input signal

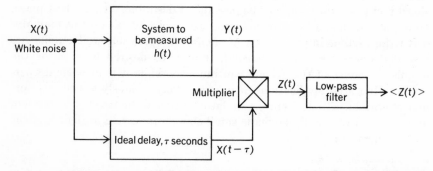

Figure 7-5 Method for measuring the impulse response of a linear system.

$X(t)$ is a sample function from a random process whose bandwidth is large compared to the bandwidth of the system to be measured. In practice, a bandwidth ratio of 10 to 1 gives very good results. For purposes of analysis this input will be assumed to be white.

In addition to being applied to the system under test, this input is also delayed by τ seconds. Depending upon the frequency range of $X(t)$ and the required value of τ, this delay may be a length of transmission line, a lumped-constant approximation to the transmission line, or a tape recorder with record and playback heads displaced enough to yield the desired delay. The output of the delay unit is simply $X(t - \tau)$.

The system output $Y(t)$ and the delay unit output are then multiplied to form $Z(t) = X(t - \tau)Y(t)$, which is then passed through a low-pass filter. If the bandwidth of the low-pass filter is sufficiently small, its output will be mostly just the dc component of $Z(t)$, with a small random component added to it. For an ergodic input process, $Z(t)$ will be ergodic[3] and the dc component of $Z(t)$ (that is, its time average) will be the same as its expected value. Thus,

$$\langle Z(t) \rangle \simeq E[Z(t)] = E[Y(t)X(t - \tau)] = R_{XY}(\tau) \tag{7-33}$$

since in the stationary case

$$E[Y(t)X(t - \tau)] = E[X(t)Y(t + \tau)] = R_{XY}(\tau) \tag{7-34}$$

But from (7-31), it is seen that

$$\langle Z(t) \rangle \simeq S_0 h(\tau) \qquad \tau \geq 0$$
$$\simeq 0 \qquad \tau < 0$$

Hence, the dc component at the output of the low-pass filter is proportional to the impulse response evaluated at the τ determined by the delay. If τ can be changed, then the complete impulse response of the system can be measured.

At first thought, this method of measuring the impulse response may seem like the hard way to solve an easy problem; it should be much easier simply to apply an impulse (or a reasonable approximation thereto) and observe the output. However, there are at least two reasons why this direct procedure may not be possible or desirable. In the first place, an impulse with sufficient area to produce an observable output may also drive the system into a region of *nonlinear* operation well outside its intended operating range. Second, it may be desired to monitor the impulse response of the system continuously while it is in normal operation. Repeated applications of impulses may seriously affect this normal operation. In the cross-correlation method, however, the random input signal can usually be made small enough to have a negligible effect on the operation.

---

[3] This is true for a time-invariant system and a fixed delay τ.

Some practical engineering situations in which this method has been successfully used include automatic control systems, chemical process control, and measurement of aircraft characteristics in flight.   One of the more exotic applications is the continuous monitoring of the impulse response of a nuclear reactor in order to observe how close it is to being critical—that is, unstable.   It is also being used to measure the dynamic response of large buildings to earth tremors or wind gusts.

Exercise 7-5

For the system and white-noise input described in Exercise 7-4 at the end of Section 7-4, evaluate both cross-correlation functions between input and output at the same values of τ.

Answer: 0, 0, 6, 0, 3, 6

7-6   Examples of Time-Domain System Analysis

A simple RC circuit responding to a random input having an exponential autocorrelation function was analyzed in Section 7-4 and was found to involve an appreciable amount of labor.   Actually, systems and inputs such as these are usually handled more conveniently by the frequency-domain methods discussed later in this chapter.   Hence, it seems desirable to look at some situation in which time-domain methods are easier. These situations occur when the impulse response and autocorrelation function have a simple form over a finite time interval.

The system chosen for this example is the *finite-time integrator*, whose impulse response is shown in Figure 7-6(a).   The input will be assumed

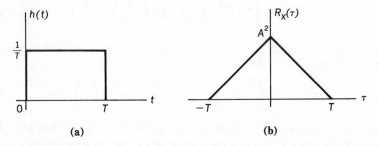

(a)   (b)

Figure 7-6   (a) Impulse response of finite-time integrator and (b) input autocorrelation function.

to have an autocorrelation function of the form shown in Figure 7-6(b). This autocorrelation function might come from the random binary process discussed in Section 5-2, for example.

For the particular input specified, the output of the finite-time integrator will have zero mean, since $\bar{X}$ is zero. In the more general case, however, the mean value of the output would be, from (7-7),

$$\bar{Y} = \bar{X} \int_0^T \frac{1}{T} \, dt = \bar{X} \tag{7-35}$$

Since the input process is not white, (7-10) must be used to determine the mean-square value of the output. Thus,

$$\overline{Y^2} = \int_0^T d\lambda_1 \int_0^T R_X(\lambda_2 - \lambda_1) \left[\frac{1}{T}\right]^2 d\lambda_2 \tag{7-36}$$

As an aid in evaluating this integral, it is helpful to sketch the integrand as shown in Figure 7-7 and note that the mean-square value is just the

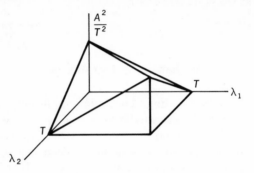

Figure 7-7 Integrand of (7-36).

volume of the region indicated. Since this volume is composed of the volumes of two right pyramids, each having a base of A^2/T^2 by $\sqrt{2}\,T$ and an altitude of $T/\sqrt{2}$, the total volume is seen to be

$$\overline{Y^2} = 2 \left\{\frac{1}{3}\right\} \left(\frac{A^2}{T^2}\right) (\sqrt{2}\,T) \left(\frac{T}{\sqrt{2}}\right) = \frac{2}{3} A^2 \tag{7-37}$$

It is also possible to obtain the autocorrelation function of the output by using (7-17). Thus,

$$R_Y(\tau) = \int_0^T d\lambda_1 \int_0^T R_X(\lambda_2 - \lambda_1 - \tau) \left(\frac{1}{T}\right)^2 d\lambda_2 \tag{7-38}$$

It is left as an exercise for the reader to show that this has the shape shown in Figure 7-8 and is composed of segments of quadratics.

It may be noted that the results become even simpler when the input random process can be treated as if it were white noise. Thus, using the special case derived in (7-12), the mean-square value of the output would be

$$\overline{Y^2} = S_0 \int_0^T \left(\frac{1}{T}\right)^2 d\lambda = \frac{S_0}{T} \tag{7-39}$$

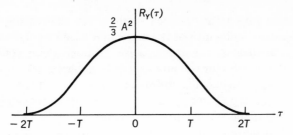

Figure 7-8   Autocorrelation function of the output of the finite-time integrator.

where S_0 is the spectral density of the input white noise.   Furthermore, from the special case derived in (7-18), the output autocorrelation function can be sketched by inspection, as shown in Figure 7-9, since it is

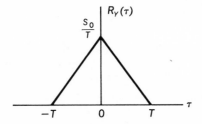

Figure 7-9   Output autocorrelation function with white-noise input.

just the convolution of the system impulse response with itself.   Note that this result indicates another way in which a random process having a triangular autocorrelation function might be generated.

The second example will utilize the result of (7-14) to determine the amount of filtering required to make a good measurement of a small dc voltage in the presence of a large noise signal.   Such a situation might arise in any system that attempts to measure the cross-correlation between two signals that are only slightly correlated.   Specifically, it will be assumed that the signal appearing at the input to the RC circuit of Figure 7-2 has the form

$$X(t) = A + N(t)$$

where the noise $N(t)$ has an autocorrelation function of

$$R_N(\tau) = 10\epsilon^{-1000|\tau|}$$

It is desired to measure A with an rms error of 1 percent when A itself is on the order of 1 and it is necessary to determine the time-constant of the RC filter required to achieve this degree of accuracy.

Although an exact solution could be obtained by using the results of the exact analysis that culminated in (7-24), this approach is needlessly complicated.   If it is recognized that the variance of the noise at the output of the filter must be very much smaller than that at the input, then it is clear that the bandwidth of the filter must also be very much smaller than the bandwidth of the input noise. Under these conditions the white-noise assumption for the input must be a very good approximation.

The first step in using this approximation is to find the spectral density of the noise in the vicinity of $\omega = 0$, since only frequency components in this region will be passed by the RC filter. Although this spectral density could be obtained directly by analogy to (7-22), the more general approach will be employed here. It was shown in (6-39) that the spectral density is related to the autocorrelation function by

$$S_N(\omega) = \int_{-\infty}^{\infty} R_N(\tau) \epsilon^{-j\omega\tau}\, d\tau$$

At $\omega = 0$, this becomes

$$S_N(0) = \int_{-\infty}^{\infty} R_N(\tau)\, d\tau = 2\int_{0}^{\infty} R_N(\tau)\, d\tau \tag{7-40}$$

Hence, the spectral density of the assumed white-noise input would be this same value, that is, $S_N = S_N(0)$. Note that (7-40) is a general result that does not depend upon the form of the autocorrelation function. In the particular case being considered here, it follows that

$$S_N = 2(10) \int_{0}^{\infty} \epsilon^{-1000\tau}\, d\tau = \frac{20}{1000} = 0.02$$

From (7-14) it is seen that the mean-square value of the filter output, $N_o(t)$, will be

$$\overline{N_o{}^2} = \frac{bS_N}{2} = \frac{b(0.02)}{2} = 0.01b$$

In order to achieve the desired accuracy of 1 percent it is necessary that

$$\sqrt{\overline{N_o{}^2}} \leq (0.01)(1.0)$$

when A is 1.0, since the dc gain of this filter is unity. Thus,

$$\overline{N_o{}^2} = 0.01b \leq 10^{-4}$$

so that

$$b \leq 10^{-2}$$

Since $b = 1/RC$, it follows that

$$RC \geq 10^2$$

in order to obtain the desired accuracy.

Exercise 7-6

Find the value of the autocorrelation function shown in Figure 7-8 at $\tau =$ (a) $T/2$, (b) T and (c) $3T/2$, if $A^2 = 12$.

Answer: 4, 7, 1

7-7   Analysis in the Frequency Domain

The most common method of representing linear systems in the frequency domain is in terms of the system function $H(\omega)$ or the transfer function $H(s)$, which are the Fourier and Laplace transforms, respectively, of the system impulse response. If the input to a system is $x(t)$ and the output $y(t)$, then the Fourier transforms of these quantities are related by

$$Y(\omega) = X(\omega)H(\omega) \qquad \text{(7-41)}$$

and the Laplace transforms are related by

$$Y(s) = X(s)H(s) \qquad \text{(7-42)}$$

provided the transforms exist. Neither of these forms is suitable when $X(t)$ is a sample function from a stationary random process. As discussed in Section 6-1, the Fourier transform of a sample function from a stationary random process generally never exists. In the case of the one-sided Laplace transform the input-output relationship is defined only for time functions existing for $t > 0$, and such time functions can never be sample functions from a stationary random process.

One approach to this problem is to make use of the spectral density of the process and to carry out the analysis using a truncated sample function in which the limit $T \to \infty$ is not taken until after the averaging operations are carried out. This procedure is valid and leads to correct results. There is, however, a much simpler procedure that can be used. In Section 6-6 it was shown that the spectral density of a stationary random process is the Fourier transform of the autocorrelation function of the process. Therefore, using the results we have already obtained for the correlation function of the output of a linear time-invariant system, we can obtain the corresponding results for the spectral density by carrying out the required transformations. When the basic relationship has been obtained, it will be seen that there is a close analogy between computations involving nonrandom signals and those involving random signals.

7-8   Spectral Density at the System Output

The spectral density of a process is a measure of how the average power of the process is distributed with respect to frequency. No information regarding the phases of the various frequency components is contained in the spectral density. The relationship between the spectral density $S_X(\omega)$ and the autocorrelation function $R_X(\tau)$ for a stationary process was shown to be

$$S_X(\omega) = \mathfrak{F}\{R_X(\tau)\} \qquad \text{(7-43)}$$

Using this relationship and (7-17), which relates the output correlation function $R_Y(\tau)$ to the input correlation function $R_X(\tau)$ by means of the system impulse response, we have

$$R_Y(\tau) = \int_0^\infty d\lambda_1 \int_0^\infty R_X(\lambda_2 - \lambda_1 - \tau) h(\lambda_1) h(\lambda_2) \, d\lambda_2$$

$$S_Y(\omega) = \mathcal{F}\{R_Y(\tau)\}$$

$$= \int_{-\infty}^\infty \left[\int_0^\infty d\lambda_1 \int_0^\infty R_X(\lambda_2 - \lambda_1 - \tau) h(\lambda_1) h(\lambda_2) \, d\lambda_2 \right] \epsilon^{-j\omega\tau} \, d\tau$$

Interchanging the order of integration and carrying out the indicated operations gives

$$S_Y(\omega) = \int_0^\infty d\lambda_1 \int_0^\infty h(\lambda_1) h(\lambda_2) \, d\lambda_2 \int_{-\infty}^\infty R_X(\lambda_2 - \lambda_1 - \tau) \epsilon^{-j\omega\tau} \, d\tau$$

$$= \int_0^\infty d\lambda_1 \int_0^\infty h(\lambda_1) h(\lambda_2) S_X(\omega) \epsilon^{-j\omega(\lambda_2 - \lambda_1)} \, d\lambda_2$$

$$= S_X(\omega) \int_0^\infty h(\lambda_1) \epsilon^{j\omega\lambda_1} \, d\lambda_1 \int_0^\infty h(\lambda_2) \epsilon^{-j\omega\lambda_2} \, d\lambda_2 \qquad \text{(7-44)}$$

$$= S_X(\omega) H(-\omega) H(\omega)$$

$$= S_X(\omega) |H(\omega)|^2$$

In arriving at (7-44) use was made of the property that $R_X(-\tau) = R_X(\tau)$.

From (7-44) it is seen that the output spectral density is related to the input spectral density by the power transfer function, $|H(\omega)|^2$. This result can also be expressed in terms of the complex frequency s as

$$S_Y(s) = S_X(s) H(s) H(-s) \qquad \text{(7-45)}$$

where $S_Y(s)$ and $S_X(s)$ are obtained from $S_Y(\omega)$ and $S_X(\omega)$ by substituting $-s^2 = \omega^2$, and where $H(s)$ is obtained from $H(\omega)$ by substituting $s = j\omega$. It is this form that will be used in further discussions of frequency analysis methods.

It is clear from (7-45) that the quantity $H(s)H(-s)$ plays the same role in relating input and output spectral densities as $H(s)$ does in relating input and output transforms. This similarity makes the use of frequency-domain techniques for systems with rational transfer functions very convenient when the input is a sample function from a *stationary random process*. However, this same technique is *not* always applicable when the input process is *nonstationary*, even though the definition for the spectral density of such processes is the same as we have employed. A detailed study of this matter is beyond the scope of the present discussion but the reader would do well to question any application of (7-45) for nonstationary processes.

Since the spectral density of the system has now been obtained, it is a simple matter to determine the mean-square value of the output. This is simply

$$\overline{Y^2} = \frac{1}{2\pi j} \int_{-j\infty}^{j\infty} H(s) H(-s) S_X(s) \, ds \qquad \text{(7-46)}$$

and may be evaluated by either of the methods discussed in Section 6-5.

In order to illustrate some of the methods, consider the RC circuit shown in Figure 7-10 and assume that its input is a sample function

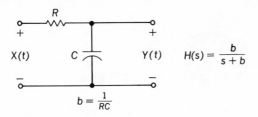

Figure 7-10 A simple RC circuit.

from a white-noise process having a spectral density of S_0. The spectral density at the output is simply

$$S_Y(s) = \frac{b}{s+b} \cdot \frac{b}{-s+b} \cdot S_0 = \frac{-b^2 S_0}{s^2 - b^2} \tag{7-47}$$

The mean-square value of the output can be obtained by using the integral I_1, tabulated in Table 6-1, Section 6-5. In order to do this it is convenient to write (7-47) as:

$$S_Y(s) = \frac{(b\sqrt{S_0})(b\sqrt{S_0})}{(s+b)(-s+b)}$$

from which it is clear that $n = 1$, and

$$c(s) = b\sqrt{S_0} = c_0$$
$$d(s) = s + b$$

Thus

$$d_0 = b$$
$$d_1 = 1$$

and

$$\overline{Y^2} = I_1 = \frac{c_0^2}{2d_0 d_1} = \frac{b^2 S_0}{2b} = \frac{bS_0}{2} \tag{7-48}$$

As a slightly more complicated example, let the input spectral density be

$$S_X(s) = \frac{-\beta^2 S_0}{s^2 - \beta^2} \tag{7-49}$$

This spectral density, which corresponds to the autocorrelation function used in Section 7-4, has been selected so that its value at zero frequency is S_0.   The spectral density at the output of the RC circuit is now

$$S_Y(s) = \frac{b}{s + b} \cdot \frac{b}{-s + b} \cdot \frac{-\beta^2 S_0}{s^2 - \beta^2}$$

$$= \frac{b^2 \beta^2 S_0}{(s^2 - b^2)(s^2 - \beta^2)} \tag{7-50}$$

The mean-square value for this output will be evaluated by using the integral I_2 tabulated in Table 6-1.   Thus,

$$S_X(s) = \frac{c(s)c(-s)}{d(s)\,d(-s)} = \frac{(b\beta \sqrt{S_0})(b\beta \sqrt{S_0})}{[s^2 + (b + \beta)s + b\beta][s^2 - (b + \beta)s + b\beta]} \tag{7-51}$$

it is clear that $n = 2$, and

$$c_0 = b\beta \sqrt{S_0}$$
$$c_1 = 0$$
$$d_0 = b\beta$$
$$d_1 = b + \beta$$
$$d_2 = 1$$

Hence,

$$\overline{Y^2} = I_2 = \frac{c_0{}^2 d_2 + c_1{}^2 d_0}{2 d_0 d_1 d_2} = \frac{b^2 \beta^2 S_0}{2 b\beta(b + \beta)} = \frac{b\beta S_0}{2(b + \beta)} \tag{7-52}$$

since $c_1 = 0$.

It is also of interest to look once again at the results when the input random process has a bandwidth much greater than the system bandwidth; that is, when $\beta \gg b$.   From (7-50) it is clear that

$$S_Y(s) = \frac{-b^2 S_0}{(s^2 - b^2)(1 - s^2/\beta^2)} \tag{7-53}$$

and as β becomes large this spectral density approaches that for the white-input-noise case given by (7-47).   In the case of the mean-square value, (7-52) may be written as

$$\overline{Y^2} = \frac{b S_0}{2(1 + b/\beta)} \tag{7-54}$$

which approaches the white-noise result of (7-48) when β is large.

Comparison of the foregoing examples with similar ones employing time-domain methods should make it evident that when the input spectral density and the system transfer function are rational, frequency-domain methods are usually simpler.   In fact, the more complicated the system, the greater the advantage of such methods.   When either the input spectral density or the system transfer function is not rational, this conclusion may not hold.

Exercise 7-8

If the input to the RC circuit of Figure 7-10 is band-limited white noise with a spectral density of 1.0 V²/Hz and a bandwidth of 50 Hz, find the mean-square value of the output when $b = 100\pi$.

Answer: 25π

7-9   Cross-spectral Densities Between Input and Output

The cross-spectral densities between a system input and output are not widely used, but it is well to be aware of their existence.   The derivation of these quantities would follow the same general pattern as shown above, but only the end results are quoted here.   Specifically, they are

$$S_{XY}(s) = H(s)S_X(s) \qquad\qquad \textbf{(7-55)}$$

and

$$S_{YX}(s) = H(-s)S_X(s) \qquad\qquad \textbf{(7-56)}$$

7-10   Examples of Frequency-Domain Analysis

Frequency-domain methods tend to be most useful when dealing with conventional filters and random processes that have rational spectral densities.   However, it is often possible to make the calculations even simpler, without introducing much error, by idealizing the filter characteristics and assuming the input processes to be white.   An important concept in doing this is that of *equivalent-noise bandwidth.*

The equivalent-noise bandwidth, B, of a system is defined to be the bandwidth of an ideal filter that has the same maximum gain and the same mean-square value at its output as the actual system when the input is white noise.   This concept is illustrated in Figure 7-11 for both low-pass and band-pass systems.   It is clear that the rectangular power transfer function of the ideal filter must have the same *area* as the power

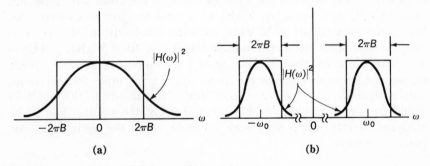

Figure 7-11   Equivalent-noise bandwidth of systems: (a) low-pass system and (b) band-pass system.

transfer function of the actual system if they are to produce the same mean-square outputs with the same white-noise input. Thus, in the low-pass case, the equivalent-noise bandwidth is given by

$$B = \frac{1}{4\pi|H(0)|^2} \int_{-\infty}^{\infty} |H(\omega)|^2 \, d\omega = \frac{1}{4\pi j|H(0)|^2} \int_{-j\infty}^{j\infty} H(s)H(-s) \, ds \quad \text{Hz}$$

(7-57)

If the input to the system is white noise with a spectral density of S_0, the mean-square value of the output is given by

$$\overline{Y^2} = 2S_0 B|H(0)|^2$$

(7-58)

In the band-pass case, $|H(0)|^2$ is replaced by $|H(\omega_0)|^2$ in both (7-57) and (7-58).

As a simple illustration of the calculation of equivalent noise bandwidth, consider the RC circuit of Figure 7-10. Since the integral of (7-57) has already been evaluated in obtaining the mean-square value of (7-48), it is easiest to use this result and (7-58). Thus,

$$\overline{Y^2} = \frac{bS_0}{2} = 2S_0 B|H(0)|^2$$

Since $|H(0)|^2 = 1$, it follows that

$$B = \frac{b}{4} = \frac{1}{4RC}$$

(7-59)

One advantage of using equivalent-noise bandwidth is that it becomes possible to describe the noise response of even very complicated systems with only two numbers, B and $|H(\omega_0)|$.

Furthermore, these numbers can be measured quite readily in an experimental system. For example, suppose that a receiver in a communication system is measured and found to have a voltage gain of 10^6 at the frequency to which it is tuned and an equivalent-noise bandwidth of 10 kHz. The noise at the input to this receiver, from shot noise and thermal agitation, has a bandwidth of several hundred megahertz and, hence, can be assumed to be white over the bandwidth of the receiver. Suppose this noise has a spectral density of 2×10^{-20} V^2/Hz. (This is a realistic value for the input circuit of a high-quality receiver.) What should the effective value of the input signal be in order to achieve an output signal-to-noise power ratio of 100? The answer to this question would be very difficult to find if every stage of the receiver had to be analyzed exactly. It is very easy, however, using the equivalent-noise bandwidth since

$$(S/N)_0 = \frac{|H(\omega_0)|^2 \overline{X^2}}{2N_0 B|H(\omega_0)|^2} = \frac{\overline{X^2}}{2N_0 B}$$

(7-60)

if $\overline{X^2}$ is the mean-square value of the input signal and N_0 is the spectral density of the input noise.   Thus,

$$\frac{\overline{X^2}}{2N_0B} = 100$$

and

$$\overline{X^2} = 2N_0B(100) = 2(2 \times 10^{-20})(10^4)(100)$$
$$= 4 \times 10^{-14}$$

from which

$$\sqrt{\overline{X^2}} = 2 \times 10^{-7} \text{ V}$$

is the effective signal voltage being sought.   Note that the actual value of receiver gain, although specified, was not needed to find the output signal-to-noise ratio.

It should be emphasized that the equivalent-noise bandwidth is useful only when one is justified in assuming that the spectral density of the input random process is white.   If the input spectral density changes appreciably over the range of frequencies passed by the system, then significant errors can result from employing the concept.

The final example of frequency-domain analysis will consider the feedback system shown in Figure 7-12.   This system might be a control

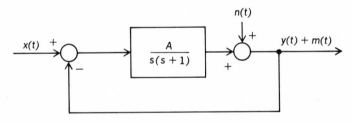

Figure 7-12 An automatic control system.

system for positioning a radar antenna, in which $x(t)$ is the input control angle (assumed to be random since target position is unknown in advance) and $y(t)$ is the angular position of the antenna in response to this voltage.   The disturbance $n(t)$ might represent the effects of wind blowing on the antenna, thus producing random perturbations on the angular position.   The transfer function of the amplifier and motor within the feedback loop is

$$H(s) = \frac{A}{s(s+1)}$$

The transfer function relating $X(s) = \mathcal{L}[x(t)]$ and $Y(s) = \mathcal{L}[y(t)]$ can be obtained by letting $n(t) = 0$ and noting that

$$Y(s) = H(s)[X(s) - Y(s)]$$

since the input to the amplifier is the difference between the input control signal and the system output.   Hence,

$$H_c(s) = \frac{Y(s)}{X(s)} = \frac{H(s)}{1 + H(s)}$$

$$= \frac{A}{s^2 + s + A}$$

(7-61)

If the spectral density of the input control signal (now considered to be a sample function from a random process) is

$$S_X(s) = \frac{-2}{s^2 - 1}$$

then the spectral density of the output is

$$S_Y(s) = S_X(s)H_c(s)H_c(-s)$$

$$= \frac{-2A^2}{(s^2 - 1)(s^2 + s + A)(s^2 - s + A)}$$

(7-62)

The mean-square value of the output is given by

$$\overline{Y^2} = \frac{2A^2}{2\pi j} \int_{-j\infty}^{j\infty} \frac{ds}{[s^3 + 2s^2 + (A + 1)s + A][-s^3 + 2s^2 - (A + 1)s + A]}$$

$$= 2A^2 I_3$$

in which

$$c_0 = 1, \qquad c_1 = 0, \qquad c_2 = 0$$
$$d_0 = A, \qquad d_1 = A + 1, \qquad d_2 = 2, \qquad d_3 = 1$$

From Table 6-1, this becomes

$$\overline{Y^2} = \frac{2A}{A + 2}$$

(7-63)

The transfer function relating $N(s) = \mathcal{L}[n(t)]$ and $M(s) = \mathcal{L}[m(t)]$ is not the same as (7-61) because the disturbance enters the system at a different point.   It is apparent, however, that

$$M(s) = N(s) - H(s)M(s)$$

from which

$$H_n(s) = \frac{M(s)}{N(s)} = \frac{1}{1 + H(s)}$$

$$= \frac{s(s + 1)}{s^2 + s + A}$$

(7-64)

Let the interfering noise have a spectral density of

$$S_N(s) = \delta(s) - \frac{1}{s^2 - 0.25}$$

This corresponds to an input disturbance that has an average value as well as a random variation. The spectral density of the output disturbance becomes

$$
\begin{aligned}
S_M(s) &= S_N(s)H_n(s)H_n(-s) \\
&= \left[\delta(s) - \frac{1}{s^2 - 0.25} \right] \frac{s^2(s^2 - 1)}{(s^2 + s + A)(s^2 - s + A)}
\end{aligned}
\qquad \text{(7-65)}
$$

The mean-square value of the output disturbance comes from

$$\overline{M^2} = \frac{1}{2\pi j} \int_{-j\infty}^{j\infty} \left[\delta(s) - \frac{1}{s^2 - 0.25} \right] \left[\frac{s^2(s^2 - 1)}{(s^2 + s + A)(s^2 - s + A)} \right] ds$$

Since the integrand vanishes at $s = 0$, the integral over $\delta(s)$ does not contribute anything to the mean-square value. The remaining terms are

$$
\begin{aligned}
\overline{M^2} &= \frac{1}{2\pi j} \int_{-j\infty}^{j\infty} \frac{s(s+1)(-s)(-s+1)}{[s^3 + 1.5s^2 + (A+0.5)s + 0.5A]} \, ds \\
&\qquad\qquad\qquad\qquad\quad [-s^3 + 1.5s^2 - (A+0.5)s + 0.5A] \\
&= I_3
\end{aligned}
$$

The constants required for Table 6-1 are

$$
\begin{array}{llll}
c_0 = 0 & c_1 = 1 & c_2 = 1 & \\
d_0 = 0.5A & d_1 = (A+0.5) & d_2 = 1.5 & d_3 = 1
\end{array}
$$

and the mean-square value becomes

$$\overline{M^2} = \frac{A + 1.5}{2A + 1.5} \qquad \text{(7-66)}$$

The amplifier gain A has not been specified in the example in order that the effects of changing this gain can be made evident. It is clear from (7-63) and (7-66) that the desired signal mean-square value increases with larger values of A while the undesired noise mean-square value decreases. Thus, one would expect that large values of A would be desirable if output signal-to-noise ratio is the important criterion. In actuality, the dynamic response of the system to rapid input changes may be more important and this would limit the value of A that can be used.

Exercise 7-10

For the RC circuit of Figure 7-10 as shown on preceding page 185, find the relationship between the equivalent-noise bandwidth, B, and the half-power bandwidth, $B_{1/2} = b/2\pi$ Hz.

Answer: $B = \dfrac{\pi}{2} B_{1/2}$

■ Problems

7-1 A white-noise process having a spectral density of S_0 V^2/Hz is applied to the circuit shown.

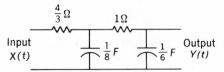

a. Find the impulse response of the network.
b. Find the dc gain of the network.
c. Find the mean-square value of the output.

7-2 A *finite-time integrator* can be represented by the block diagram shown.

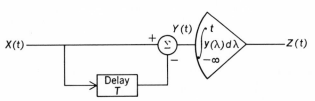

a. Find the impulse response of the system.
b. Find the dc gain of the system.
c. Is this system stable?
d. Using time-domain techniques find the mean-square value of the output if white noise having a spectral density of S_0 V^2/Hz is applied to the input.

7-3 A sample function from a random process is a voltage of the form

$$X(t) = X_0 + \cos(2\pi t + \theta)$$

where X_0 is a random variable uniformly distributed from 0 to 1, θ is a random variable that is independent of X_0 and is uniformly distributed from 0 to 2π. If this process is applied to the finite-time integrator of Problem 7-2, find the mean and variance of the output.

7-4 White noise having a spectral density of S_0 V^2/Hz is applied to the circuit shown on the following page. Using time-domain methods,

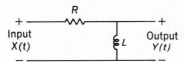

find the autocorrelation function of the output.

7-5 Repeat Problem 7-4 with an input having an autocorrelation function of the form

$$R_X(\tau) = \frac{\beta S_0}{2} \epsilon^{-\beta|\tau|}$$

7-6 The random process described in Problem 7-3 is applied to the circuit shown.   Find the autocorrelation function of the output signal.

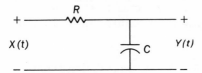

7-7 For the system and white-noise input of Problem 7-4, find both cross-correlation functions between input and output.

7-8 For the system of Problem 7-4 and the input given in Problem 7-5, find both cross-correlations between the input and output.

7-9 A stationary white-noise process, $X(t)$ having a spectral density of S_0 V^2/Hz is applied to the circuit shown.   Using time-domain techniques, find the cross-correlation function $R_{YZ}(\tau)$ for all τ.

7-10 A random process has sample functions of the form

$$X(t) = A + N(t)$$

where A is a signal of constant magnitude and $N(t)$ is a random noise having an autocorrelation function of

$$R_N(\tau) = \epsilon^{-|\tau|}$$

It is desired to separate the signal from the noise by passing $X(t)$ through a single section low-pass RC filter.   If the signal-to-noise power ratio is defined as

$$\left[\frac{S}{N}\right]_0 = \frac{\text{output signal power}}{\text{output variance}} = \frac{(\bar{Y})^2}{\overline{Y^2} - (\bar{Y})^2}$$

where $Y(t)$ is the output of the low-pass filter, find the filter time constant required to obtain a signal-to-noise ratio of 30 dB if the true value of A is 0.1 V.

7-11   Solve Problem 7-1 using frequency-domain methods.

7-12   Solve Problem 7-2(d) using frequency-domain methods.

7-13   Find the spectral density and autocorrelation function of the output in Problem 7-4 using frequency-domain methods.

7-14   A tuned amplifier has its maximum gain of 30 dB at a frequency of 30 MHz and a halfpower bandwidth of 1 MHz.   The response curve has a shape equivalent to that of a single stage parallel RLC circuit. A white-noise source is connected to the input of the amplifier and the rms value of the output is measured to be 10 volts.   Find the spectral density of the input signal.

7-15   Find and sketch the autocorrelation function of the output signal of an ideal bandpass filter having a center frequency of f_0, a bandwidth of W, and an input of white noise with spectral density of N_0.

7-16   It has been proposed to measure the range to a reflecting object by transmitting a bandlimited white-noise signal at a carrier frequency of f_0 and then adding the received signal to the transmitted signal and measuring the spectral density of the sum.   Periodicities in the ampli-

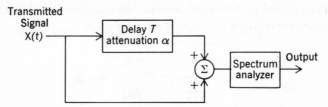

tude of the spectral density are related to the range.   Using the system model shown and assuming that α^2 is negligible compared to α, investigate the possibility of this approach.   What effect that would adversely affect the measurement has been omitted in the system model?

7-17   It is frequently useful to approximate the shape of a filter by a Gaussian function of frequency.   Determine the standard deviation of a Gaussian shaped lowpass filter that has a maximum gain of unity and a halfpower bandwidth of W Hz.   Find the equivalent noise bandwidth of a Gaussian shaped filter in terms of its halfpower bandwidth and in terms of its standard deviation.

7-18   The thermal agitation noise generated by a resistance can be closely approximated as white noise having a spectral density of $2kTR\ V^2/\text{Hz}$, where $k = 1.37 \times 10^{-23}\ W\text{-}s/°K$ is the Boltzmann constant, T is the absolute temperature in degrees Kelvin, and R is the resistance in ohms. Any physical resistance in an amplifier is paralleled by a capacitance, so that the equivalent circuit is as shown.

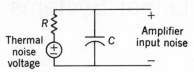

a. Calculate the mean-square value of the amplifier input noise and show that it is independent of R.
b. Explain this result on a physical basis.
c. Show that the maximum noise power available (that is, with a matched load) from a resistance is kTB watts, where B is the equivalent noise bandwidth over which the power is measured.

7-19 Any signal at the input of an amplifier is always accompanied by noise. The minimum noise theoretically possible is the thermal noise present in the resistive component of the input impedance as described in Problem 7-18. In general the amplifier will add additional noise in the process of amplifying the signal. The amount of noise is measured in terms of the deterioration of the signal-to-noise ratio of the signal when it is passed through the amplifier. A common method of specifying this characteristic of an amplifier is in terms of a noise figure F, defined as

$$F = \frac{\text{input signal-to-noise power ratio}}{\text{output signal-to-noise power ratio}}$$

a. Using the above definition show that the overall noise figure for two cascaded amplifiers is $F = F_1 + (F_2 - 1)/G_1$ where the individual amplifiers have power gains of G_1 and G_2 and noise figures of F_1 and F_2 respectively.
b. A particular wide-band video amplifier has a single time constant roll-off with a half-power bandwidth of 100 MHz, a gain of 100 dB, a noise figure of 13 dB, and input and output impedances of 300 Ω. Find the rms output noise voltage when the input signal is zero.
c. Find the amplitude of the input sine wave required to give an output signal-to-noise power ratio of 10 dB.

■REFERENCES

See References for Chapter 1, particularly Davenport and Root, Lanning and Battin, Papoulis, and Thomas.

8
Optimum Linear Systems

8-1   Introduction

It has been pointed out previously that almost any practical system has some sort of random disturbance introduced into it in addition to the desired signal.   The presence of this random disturbance means that the system output is never quite what it should be, and may deviate very considerably from its desired value.   When this occurs, it is natural to ask if the system can be modified in any way to reduce the effects of the disturbances.   Usually it turns out that it is possible to select a system impulse response or transfer function that minimizes some attribute of the output disturbance.   Such a system is said to be *optimum*.

The study of optimum systems for various types of desired signals and various types of noise disturbance is very involved because of the many different situations that can be specified.   The literature on the subject is quite extensive and the methods used to determine the optimum system are quite general, quite powerful, and quite beyond the scope of the present discussion.   Nevertheless, it is desirable to introduce some of the terminology and a few of the basic concepts in order that the student be aware of some of the possibilities and be in a better position to read the literature.

One of the first steps in the study of optimum systems is a precise definition of what constitutes optimality. Since many different criteria of optimality exist, it is necessary to use some care in selecting an appropriate one. This problem is discussed in the following section.

After a criterion has been selected, the next step is to specify the nature of the system to be considered. Again there are many possibilities, and the ease of carrying out the optimization may be critically dependent upon the choice. Section 8-3 considers this problem briefly.

Once the optimum system has been determined, there remains the problem of evaluating its performance. In some cases this is relatively easy, while in other cases it may be more difficult than actually determining the optimum system. No general treatment of the evaluation problem will be given here; each case considered will be handled separately.

In an actual engineering problem, the final step is to decide if the optimum system can be built economically, or whether it will be necessary to approximate it. If it turns out, as it often does, that it is not possible to build the true optimum system, then it is reasonable to question the value of the optimization techniques. Strangely enough, however, it is frequently useful and desirable to carry out the optimizing exercise even though there is no intention of attempting to construct the optimum system. The reason is that the optimum performance provides a yardstick against which the performance of any actual system can be compared. Since the optimum performance cannot be exceeded, this comparison clearly indicates whether any given system needs to be improved or whether its performance is already so close to the optimum that further effort on improving it would be uneconomical. In fact, it is probably this type of comparison that provides the greatest motivation for studying optimum systems since it is only rarely that the true optimum system can actually be constructed.

8-2 Criteria of Optimality

Since there are many different criteria of optimality that might be selected, it is necessary to establish some guidelines as to what constitutes a reasonable criterion. In the first place, it is necessary that the criterion satisfy certain requirements, such as:

1. The criterion must have physical significance and not lead to a trivial result. For example, if the criterion were that of minimizing the output noise power, the obvious result would be a system having *zero* output for both signal and noise. This is clearly a trivial result. On the other hand, a criterion of minimizing the output noise power subject to the constraint of maintaining a given output signal power might be quite reasonable.

2. The criterion must lead to a unique solution. For example, the criterion that the *average error* of the output signal be zero can be satisfied by many systems, not all equally good in regard to the *variance* of the error.

3. The criterion should result in a mathematical form that is capable of being solved. This requirement turns out to be a very stringent one and is the primary reason why so few criteria have found practical application. As a consequence, the criterion is often selected primarily on this basis even though some other criterion might be more desirable in a given situation.

The choice of a criterion is often influenced by the nature of the input signal—that is, whether it is deterministic or random. The reason for this is that the purpose of the system is usually different for these two types of signals. For example, if the input signal is deterministic, then its form is known and the purpose in observing it is to determine such things as whether it is present or not, the time at which it occurs, how big it is, and so on. On the other hand, when the signal is random its form is unknown and the purpose of the system is usually to determine its form as nearly as possible. In either of these cases there are a number of criteria that might make sense. However, only one criterion for each case will be discussed here, and the one selected will be the one that is most common and most easily handled mathematically.

In the case of deterministic signals, the criterion of optimality will be to *maximize* the output signal-to-noise power ratio at some specified time. This criterion is particularly useful when the purpose of the system is to detect the presence of a signal of known shape or to measure the time at which such a signal occurs. There is some flexibility in this criterion with respect to choosing the time at which the signal-to-noise ratio is to be maximized, but reasonable choices are usually apparent from the nature of the signal.

In the case of random signals, the criterion of optimality will be to *minimize* the mean-square value of the difference between the actual system output and the actual value of the signal being observed. This criterion is particularly useful when the purpose of the system is to observe an unknown signal for purposes of measurement or control. The difference between the output of the system and the true value of the signal consists of two components. One component is the *signal error* and represents the difference between the input and output when there is no input noise. The second component is the output noise, which also represents an error in the output. The total error is the sum of these components, and the quantity to be minimized is the mean-square value of this total error.

8-3 Restrictions on the Optimum System

It is usually necessary to impose some sort of restriction on the type of system that will be permitted. The most common restriction is that the system must be *causal*,[1] since this is a fundamental requirement of physical realizability. It frequently is true that a noncausal system, which can respond to *future* values of the input, could do a better job of satisfying the chosen criterion than any physically realizable system. A noncausal system cannot be built, however, and does not provide a fair comparison with real systems, so is usually inappropriate. A possible exception to this rule arises when the data available to the system is in recorded form so that future values can, in fact, be utilized.

Another common assumption is that the system is *linear*. The major reason for this assumption is that it is usually not possible to carry out the analytical solution for the optimum nonlinear system. In many cases, particularly those involving Gaussian noise, it is possible to show that there is no nonlinear system that will do a better job than the optimum linear system. However, in the more general case, the linear system may not be the best. Nevertheless, the difficulty of determining the optimum nonlinear system is such that it is usually not feasible to hunt for it.

Once a reasonable criterion has been selected, and the system restricted to being causal and linear, then it is usually possible to find the impulse response or transfer function that optimizes the criterion. However, in some cases it may be desirable to further restrict the system to a particular form. The reason for such a restriction is usually that it guarantees a system having a given complexity (and, hence, cost) while the more general optimization may yield a system that is costly or difficult to approximate. An example of this specialized type of optimization will be considered in the next section.

8-4 Optimization by Parameter Adjustment

As suggested by the title, this method of optimization is carried out by specifying the form of the system to be used and then by finding the values of the components of that system that optimize the selected criterion. This procedure has the obvious advantage of yielding a system whose complexity can be predetermined and, hence, has its greatest application in cases in which the complexity of the system is of critical importance because of size, weight, or cost considerations. The dis-

---

[1] By *causal* we mean that the system impulse response satisfies the condition $h(t) = 0$, $t < 0$ [see equation (7-3)]. In addition, the *stability* condition of equation (7-4) is also assumed to apply.

advantage is that the performance of this type of optimum system will never be quite as good as that of a more general system whose form is not specified in advance. Any attempt to improve performance by picking a slightly more complicated system to start out with leads to analytical problems in determining the optimum values of more than one parameter (because the simultaneous equations that must be solved are seldom linear), although computer solutions are quite possible. As a practical matter, analytical solutions are usually limited to a single parameter. Two different examples will be discussed in order to illustrate the basic ideas.

As a first example, assume that the signal consists of a rectangular pulse as shown in Figure 8-1(a) and that this signal is combined with

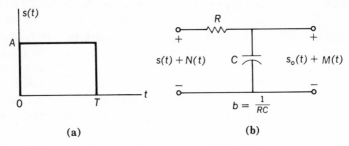

(a)   **(b)**

Figure 8-1 Signal and system for maximizing signal-to-noise ratio: **(a)** signal to be detected and **(b)** specified form of optimum system.

white noise having a spectral density of N_o. Since the form of the signal is known, the objective of the system is to detect the presence of the signal. As noted earlier, a reasonable criterion for this purpose is to find that system maximizing the output signal-to-noise power ratio at some instant of time. That is, if the output signal is $s_o(t)$, and the mean-square value of the output noise is $\overline{M^2}$, then it is desired to find the system that will maximize the ratio $s_o{}^2(t_o)/\overline{M^2}$, where t_o is the time chosen for this to be a maximum.

In the method of parameter adjustment, the form of the system is specified and in this case is assumed to be a simple RC circuit, as shown in Figure 8-1(b). The parameter to be adjusted is the time constant of the filter—or, rather, the reciprocal of this time constant. One of the first steps is to select a time t_o at which the signal-to-noise ratio is a maximum. An appropriate choice for t_o becomes apparent when the output signal component is considered. This output signal is given by

$$\begin{aligned} s_o(t) &= A[1 - \epsilon^{-bt}] & 0 \le t < T \\ &= A[1 - \epsilon^{-bT}]\epsilon^{-b(t-T)} & T \le t < \infty \end{aligned} \tag{8-1}$$

and is sketched in Figure 8-2. This result is arrived at by any of the conventional methods of system analysis. It is clear from this sketch that the output signal component has its largest value at time T. Hence,

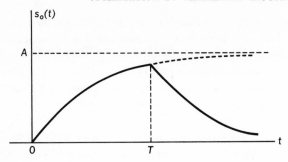

Figure 8-2 Signal component at the RC filter output.

it is reasonable to choose $t_o = T$, and thus

$$s_o(t_o) = A(1 - \epsilon^{-bT}) \tag{8-2}$$

The mean-square value of the output noise from this type of circuit has been considered several times before and has been shown to be

$$\overline{M^2} = \frac{bN_o}{2} \tag{8-3}$$

Hence, the signal-to-noise ratio to be maximized is

$$\frac{s_o{}^2(t_o)}{\overline{M^2}} = \frac{A^2(1 - \epsilon^{-bT})^2}{bN_o/2} \qquad b \geq 0 \tag{8-4}$$

Before carrying out the maximization, it is worth noting that this ratio is zero for both $b = 0$ and $b = \infty$, and that it is positive for all other positive values of b. Hence, there must be some positive value of b for which the ratio is a maximum.

In order to find the value of b that maximizes the ratio, (8-4) is differentiated with respect to b and the derivative equated to zero. Thus,

$$\frac{d[s_o{}^2(t_o)/\overline{M^2}]}{db} = \frac{2A^2}{N_o} \frac{[2b(1 - \epsilon^{-bT})T\epsilon^{-bT} - (1 - \epsilon^{-bT})^2]}{b^2} = 0 \tag{8-5}$$

This can be simplified to yield the nontrivial equation

$$2bT + 1 = \epsilon^{bT} \tag{8-6}$$

This equation is easily solved for bT by trial-and-error methods and leads to

$$bT \doteq 1.26 \tag{8-7}$$

from which the optimum time constant is

$$RC = \frac{T}{1.26} \tag{8-8}$$

This, then, is the value of time constant that should be used in the RC filter in order to maximize the signal-to-noise ratio at time T.

The next step in the procedure is to determine how good the filter actually is. This is easily done by substituting the optimum value of bT, as given by (8-7), into the signal-to-noise ratio of (8-4). When this is done, it is found that

$$\left[\frac{s_o{}^2(t_o)}{M^2}\right]_{\max} = \frac{0.814 A^2 T}{N_o} \tag{8-9}$$

It may be noted that the *energy* of the pulse is $A^2 T$, so that the maximum signal-to-noise ratio is proportional to the ratio of the signal energy to the noise spectral density. This is typical of *all* cases of maximizing signal-to-noise ratio in the presence of white noise. In a later section, it will be shown that if the form of the optimum system were not specified as it was here, but allowed to be general, the constant of proportionality would be 1.0 instead of 0.814. The reduction in signal-to-noise ratio encountered in this example may be considered as the price that must be paid for insisting on a simple filter. The loss is not serious here, but in other cases it may be appreciable.

The final step, which is frequently omitted, is to determine just how sensitive the signal-to-noise ratio is to the choice of the parameter b. This is most easily done by sketching the proportionality constant in (8-4) as a function of b. Since this constant is simply

$$K = \frac{2(1 - \epsilon^{-bT})^2}{bT} \tag{8-10}$$

the result is as shown in Figure 8-3. It is clear from this sketch that the output signal-to-noise ratio does not change rapidly with b in the vicinity

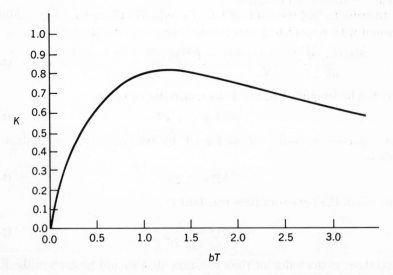

Figure 8-3 Output signal-to-noise ratio as function of the parameter bT.

of the maximum so that it is not very important to have precisely the right value of time constant in the optimum filter.

The second example of optimization by parameter adjustment will consider a random signal and employ the minimum mean-square error criterion. In this example the system will be an ideal low-pass filter, rather than a specified circuit configuration, and the parameter to be adjusted will be the bandwidth of that filter.

Assume that the signal $X(t)$ is a sample function from a random process having a spectral density of

$$S_X(\omega) = \frac{A^2}{\omega^2 + (2\pi f_a)^2} \tag{8-11}$$

Added to this signal is white noise $N(t)$ having a spectral density of N_o. These are illustrated in Figure 8-4 along with the power transfer charac-

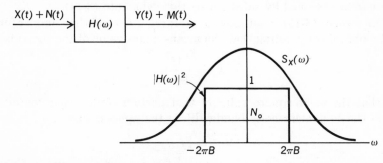

Figure 8-4 Signal and noise spectral densities, filter characteristic.

teristic of the ideal low-pass filter.

Since the filter is an ideal low-pass filter, the error in the output *signal component*, $E(t) = X(t) - Y(t)$, will be due entirely to that portion of the signal spectral density falling outside the filter pass band. Its mean-square value can be obtained by integrating the signal spectral density over the region *outside* of $\pm 2\pi B$. Because of symmetry, only one side need be evaluated and then doubled. Hence,

$$\begin{aligned} \overline{E^2} &= \frac{2}{2\pi} \int_{2\pi B}^{\infty} \frac{A^2}{\omega^2 + (2\pi f_a)^2} \, d\omega \\ &= \frac{2A^2}{4\pi^2 f_a} \left[\frac{\pi}{2} - \tan^{-1} \frac{B}{f_a} \right] \end{aligned} \tag{8-12}$$

The noise out of the filter, $M(t)$, has a mean-square value of

$$\overline{M^2} = \frac{1}{2\pi} \int_{-2\pi B}^{2\pi B} N_o \, d\omega = 2BN_o \tag{8-13}$$

The total mean-square error is the sum of these two (since signal and noise are statistically independent) and is the quantity that is to be minimized by selecting B.   Thus,

$$\overline{E^2} + \overline{M^2} = \frac{2A^2}{4\pi^2 f_a}\left[\frac{\pi}{2} - \tan^{-1}\frac{B}{f_a}\right] + 2BN_o \qquad \text{(8-14)}$$

The minimization is accomplished by differentiating (8-14) with respect to B and setting the result equal to zero.   Thus,

$$\frac{2A^2}{4\pi^2 f_a}\left[\frac{-1/f_a}{1 + (B/f_a)^2}\right] + 2N_o = 0$$

from which it follows that

$$B = \left[\frac{A^2}{4\pi^2 N_o} - f_a{}^2\right]^{1/2} \qquad \text{(8-15)}$$

is the optimum value.   The actual value of the minimum mean-square error can be obtained by substituting this value into (8-14).

The form of (8-15) is not easy to interpret.   A somewhat simpler form can be obtained by noting that the mean-square value of the signal is

$$\overline{X^2} = \frac{A^2}{4\pi f_a}$$

and that the mean-square value of that portion of the *noise* contained within the equivalent-noise bandwidth of the *signal* is just

$$\overline{N_X{}^2} = \pi f_a N_o$$

since the equivalent-noise bandwidth of the signal is $(\pi/2)f_a$.   Hence, (8-15) can be written as

$$B = f_a\left[\frac{\overline{X^2}}{\overline{N_X{}^2}} - 1\right]^{1/2} \qquad \overline{X^2} > \overline{N_X{}^2} \qquad \text{(8-16)}$$

and sketched as in Figure 8-5.

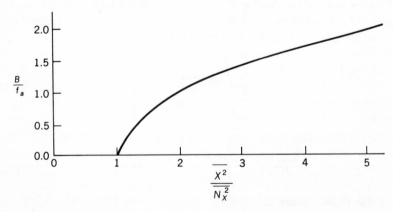

Figure 8-5   Optimum bandwidth.

In each of the examples just discussed only *one* parameter of the system was adjusted in order to optimize the desired criterion. The procedure for adjusting two or more parameters is quite similar. That is, the quantity to be maximized or minimized is differentiated with respect to *each* of the parameters to be adjusted and the derivatives set equal to zero. This yields a set of simultaneous equations that can, in principle, be solved to yield the desired parameter values. As a practical matter, however, this procedure is only rarely possible because the equations are usually nonlinear and analytical solutions are not known. Computer solutions may be obtained frequently, but there may be unresolved questions concerning uniqueness.

Exercise 8-4

In the example above it is not clear from (8-16) and Figure 8-5 just what the optimum bandwidth should be when $\overline{X^2} < \overline{N_X^2}$. Show that this optimum bandwidth is $B = 0$ by sketching (without computation) the total mean-square error of (8-14) as a function of B.

8-5   Systems That Maximize Signal-to-Noise Ratio

This section will consider systems that maximize signal-to-noise ratio at a specified time, when the form of the signal is known. The form of the system is *not* specified; the only restrictions on the system being that it must be *causal* and *linear*.

The notation is illustrated in Figure 8-6. The signal $s(t)$ is deter-

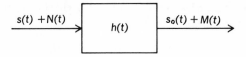

Figure 8-6   Notation for optimum filter.

ministic and assumed to be known (except, possibly, for amplitude and time of occurrence). The noise $N(t)$ is assumed to be white with a spectral density of N_o. Although the case of nonwhite noise will not be considered here (except for a brief mention at the end of this section) the same general procedure can be used for it also. The output signal-to-noise ratio is defined to be $s_o^2(t_o)/\overline{M^2}$, and the time t_o is to be selected. The objective is to find the form of $h(t)$ that will maximize this output signal-to-noise ratio.

In the first place, the output signal is given by

$$s_o(t) = \int_0^\infty h(\lambda)s(t - \lambda)\, d\lambda \tag{8-17}$$

and the mean-square value of the output noise is, for a white noise input, given by

$$\overline{M^2} = N_o \int_0^\infty h^2(\lambda) \, d\lambda \tag{8-18}$$

Hence, the signal-to-noise ratio at time t_o is

$$\frac{s_o^2(t_o)}{\overline{M^2}} = \frac{\left[\int_0^\infty h(\lambda) s(t_o - \lambda) \, d\lambda \right]^2}{N_o \int_0^\infty h^2(\lambda) \, d\lambda} \tag{8-19}$$

In order to maximize this ratio it is convenient to use the *Schwarz inequality*. This inequality states that for any two functions, say $f(t)$ and $g(t)$, that

$$\left[\int_a^b f(t)g(t) \, dt \right]^2 \leq \int_a^b f^2(t) \, dt \int_a^b g^2(t) \, dt \tag{8-20}$$

Furthermore, the *equality* holds if and only if $f(t) = kg(t)$, where k is independent of t.

Using the Schwarz inequality on (8-19) leads to

$$\frac{s_o^2(t_o)}{\overline{M^2}} \leq \frac{\int_0^\infty h^2(\lambda) \, d\lambda \int_0^\infty s^2(t_o - \lambda) \, d\lambda}{N_o \int_0^\infty h^2(\lambda) \, d\lambda} \tag{8-21}$$

From this it is clear that the maximum value of the signal-to-noise ratio occurs when the equality holds, and that this maximum value is just

$$\left[\frac{s_o^2(t_o)}{\overline{M^2}} \right]_{\text{max}} = \frac{1}{N_o} \int_0^\infty s^2(t_o - \lambda) \, d\lambda \tag{8-22}$$

since the integrals of $h^2(\lambda)$ cancel out. Furthermore, the condition that is required for the equality to hold is

$$h(\lambda) = k s(t_o - \lambda) u(\lambda) \tag{8-23}$$

Since the k is simply a gain constant that does not affect the signal-to-noise ratio, it can be set equal to any value; a convenient value is $k = 1$. The $u(\lambda)$ has been added to guarantee that the system is causal. Note that the desired impulse response is simply the signal waveform run backwards in time and delayed by t_o seconds.

The right side of (8-22) can be written in slightly different form by letting $t = t_o - \lambda$. Upon making this change of variable, the integral becomes

$$\int_0^\infty s^2(t_o - \lambda) \, d\lambda = \int_{-\infty}^{t_o} s^2(t) \, dt = \mathcal{E}(t_o) \tag{8-24}$$

and it is clear that this is simply the energy in the signal up to the time the signal-to-noise ratio is to be maximized. This signal energy is designated as $\mathcal{E}(t_o)$.

To summarize, then:

1. The output signal-to-noise ratio at time t_o is maximized by a filter whose impulse response is

$$h(t) = s(t_o - t)u(t) \qquad \text{(8-25)}$$

2. The value of the maximum signal-to-noise ratio is

$$\left[\frac{s_o{}^2(t_o)}{\overline{M^2}}\right]_{\max} = \frac{\mathcal{E}(t_o)}{N_o} \qquad \text{(8-26)}$$

where $\mathcal{E}(t_o)$ is the energy in $s(t)$ up to the time t_o.

The filter defined by (8-25) is usually referred to as a *matched filter*.

As a first example of this procedure, consider again the case of the rectangular signal pulse, as shown in Figure 8-7(a), and find the $h(t)$ that

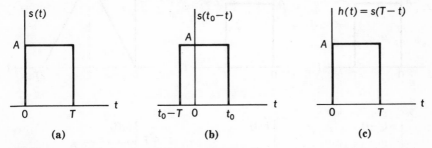

Figure 8-7 Matched filter for a rectangular pulse: (**a**) signal, (**b**) reversed, translated signal, and (**c**) optimum filter for $t_o = T$.

will maximize the signal-to-noise ratio at $t_o = T$. The reversed and translated signal is shown (for an arbitrary t_o) in Figure 8-7(b). The resulting impulse response for $t_o = T$ is shown in Figure 8-7(c) and is represented mathematically by

$$\begin{aligned} h(t) &= A \qquad 0 \le t \le T \\ &= 0 \qquad \text{elsewhere} \end{aligned} \qquad \text{(8-27)}$$

The maximum signal-to-noise ratio is given by

$$\left[\frac{s_o{}^2(t_o)}{\overline{M^2}}\right]_{\max} = \frac{\mathcal{E}(t_o)}{N_o} = \frac{A^2 T}{N_o} \qquad \text{(8-28)}$$

This result may be compared with (8-9).

In order to see the effect of changing the value of t_o, the sketches of Figure 8-8 are presented. The sketches show $s(t_o - t)$, $h(t)$, and the output signal $s_o(t)$, all for the same input $s(t)$ shown in Figure 8-7(a). It is clear from these sketches that making $t_o < T$ decreases the maximum signal-to-noise ratio because not all of the energy of the pulse is available at time t_o. On the other hand, making $t_o > T$ does not further

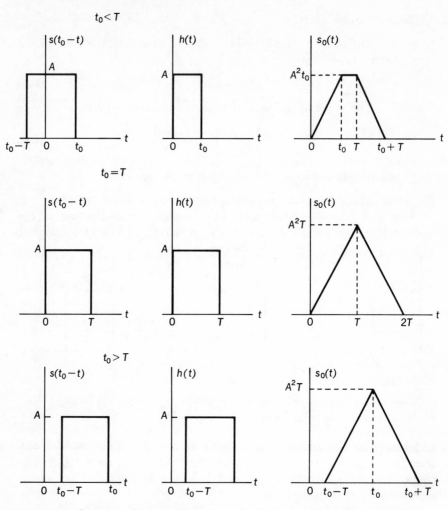

Figure 8-8 Optimum filters and responses for various values of t_o.

increase the output signal-to-noise ratio, since all of the pulse energy is available by time T. It is also clear that the signal out of the matched filter does not have the same shape as the input signal. Thus, the matched filter is *not* suitable if the objective of the filter is to recover a nearly undistorted rectangular pulse.

As a second example of matched filters, it is of interest to consider a signal having finite energy but infinite time duration. Such a signal might be

$$s(t) = A\epsilon^{-bt}u(t) \tag{8-29}$$

as shown in Figure 8-9. For some arbitrarily selected t_o, the optimum matched filter is

$$h(t) = A\epsilon^{-b(t_o-t)}u(t) \tag{8-30}$$

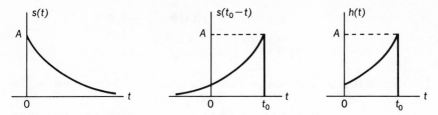

Figure 8-9   Matched filter for an exponential signal.

and is also shown.   The maximum signal-to-noise ratio depends upon t_o, since the available energy increases with t_o.   In this case it is

$$\left[\frac{s_o^2(t_o)}{\overline{M^2}}\right]_{\max} = \frac{\mathcal{E}(t_o)}{N_o} = \frac{\int_0^{t_o} A^2 \epsilon^{-2bt}\, dt}{N_o} = \frac{A^2}{2bN_o}\,[1 - \epsilon^{-2bt_o}] \qquad \textbf{(8-31)}$$

It is clear that this approaches a limiting value of $A^2/2bN_o$ as t_o is made large.   Hence, the choice of t_o is governed by how close to this limit one wants to come—remembering that larger values of t_o generally represent a more costly system.

The third and final illustration of the matched filter will consider signals having both infinite energy and infinite time duration, that is, power signals.   Any periodically repeated waveform would be an example of this type of signal.   A case of considerable interest is that of periodically repeated RF pulses such as would be used in a pulse radar system.   Figure 8-10 shows such a signal, the corresponding reversed, translated signal, and the impulse response of the matched filter.   In this sketch, t_o has been shown as containing an integral number of pulses, but this is not necessary.   Since the energy per pulse is just $\frac{1}{2}A^2t_p$, the

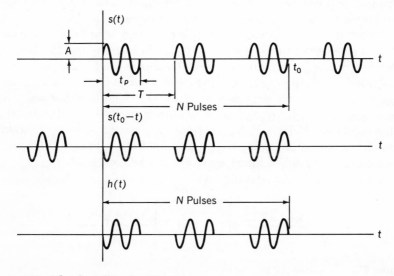

Figure 8-10   Matched filter for N pulses.

signal-to-noise ratio out of a filter matched to N such pulses is

$$\left[\frac{s_o{}^2(t_o)}{\overline{M^2}}\right]_{\text{max}} = \frac{NA^2t_p}{2N_0} \tag{8-32}$$

It is clear that this signal-to-noise ratio continues to increase as the number of pulses included in the matched filter increases. However, it becomes very difficult to build filters that are matched for very large values of N, so usually N is a number less than 10.

Although it is not intended to discuss the case of nonwhite noise in any detail, it may be noted that all that is needed in order to apply the above matched filter concepts is to precede the matched filter with a network that converts the nonwhite noise into white noise. Such a device is called a *prewhitening* filter and has a power transfer function that is the reciprocal of the noise spectral density. Of course, the prewhitening filter changes the shape of the signal so that the subsequent matched filter has to be matched to this new signal shape rather than to the original signal shape.

Exercise 8–5

For the matched filter of (8-30), find the value of t_o that would make the maximum output signal-to-noise ratio equal 0.9 of its limiting value.

Answer: $t_o = (\ln 10)/2b$

8-6 Systems That Minimize Mean-Square Error

This section considers systems that minimize the mean-square error between the total system output and the input signal component when that signal is from a stationary random process. The form of the system is not specified in advance, but it is restricted to be linear and causal.

It is convenient to use s-plane notation in carrying out this analysis, although it can be done in the time-domain as well. The notation is illustrated in Figure 8-11, in which the random input signal $X(t)$ is assumed to have a spectral density of $S_X(s)$, while the input noise $N(t)$ has a spectral density of $S_N(s)$. The output spectral densities for these components are $S_Y(s)$ and $S_M(s)$, respectively. There is no particular simplification from assuming the input noise to be white (as there was in the case of the matched filter), so it will not be done here.

The error in the signal component, produced by the system, is defined

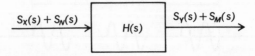

Figure 8-11 Notation for the optimum system.

as before by

$$E(t) = X(t) - Y(t)$$

and its Laplace transform is

$$F_E(s) = F_X(s) - F_Y(s) = F_X(s) - H(s)F_X(s) = F_X(s)[1 - H(s)] \quad \text{(8-33)}$$

Hence, $1 - H(s)$ is the transfer function relating the *signal error* to the *input signal*, and the mean-square value of the signal error is given by

$$\overline{E^2} = \frac{1}{2\pi j} \int_{-j\infty}^{j\infty} S_X(s)[1 - H(s)][1 - H(-s)] \, ds \quad \text{(8-34)}$$

The noise appearing at the system output is $M(t)$, and its mean-square value is

$$\overline{M^2} = \frac{1}{2\pi j} \int_{-j\infty}^{j\infty} S_N(s)H(s)H(-s) \, ds \quad \text{(8-35)}$$

The total mean-square error is $\overline{E^2} + \overline{M^2}$ (since signal and noise are statistically independent) and may be expressed as

$$\overline{E^2} + \overline{M^2} = \frac{1}{2\pi j} \int_{-j\infty}^{j\infty} \{ S_X(s)[1 - H(s)][1 - H(-s)] \\ + S_N(s)H(s)H(-s) \} \, ds \quad \text{(8-36)}$$

The objective now is to find the form of $H(s)$ that minimizes (8-36).

If there were no requirement that the system be causal, finding the optimum value of $H(s)$ would be very simple. In order to do this, rearrange the terms in (8-36) as

$$\overline{E^2} + \overline{M^2} = \frac{1}{2\pi j} \int_{-j\infty}^{j\infty} \{ [S_X(s) + S_N(s)]H(s)H(-s) \\ - S_X(s)H(s) - S_X(s)H(-s) + S_X(s) \} \, ds \quad \text{(8-37)}$$

Since $[S_X(s) + S_N(s)]$ is also a spectral density, it must have the same symmetry properties and, hence, can be factored into one factor having poles and zeros in the left half plane and another factor having the same poles and zeros in the right half plane. Thus, it can always be written as

$$S_X(s) + S_N(s) = F_i(s)F_i(-s) \quad \text{(8-38)}$$

Substituting this into (8-37), and again rearranging terms, leads to

$$\overline{E^2} + \overline{M^2} = \frac{1}{2\pi j} \int_{-j\infty}^{j\infty} \left\{ \left[F_i(s)H(s) - \frac{S_X(s)}{F_i(-s)} \right] \left[F_i(-s)H(-s) - \frac{S_X(s)}{F_i(s)} \right] \\ + \frac{S_X(s)S_N(s)}{F_i(s)F_i(-s)} \right\} \, ds \quad \text{(8-39)}$$

It may now be noted that the last term of (8-39) does not involve $H(s)$. Hence, the minimum value of $\overline{E^2} + \overline{M^2}$ will occur when the two factors in the first term of (8-39) are zero (since the product of these factors cannot be negative). This implies, therefore, that

$$H(s) = \frac{S_X(s)}{F_i(s)F_i(-s)} = \frac{S_X(s)}{S_X(s) + S_N(s)} \qquad \text{(8-40)}$$

should be the optimum transfer function. This would be true except that (8-40) is also symmetrical in the s-plane and, hence, cannot represent a causal system.

Since the $H(s)$ defined by (8-40) is not causal, the first inclination is to simply use the left half plane poles and zeros of (8-40) to define a causal system. This would appear to be analogous to eliminating the negative time portion of $s(t_o - t)$ in the matched filter of the previous section. Unfortunately, the problem is not quite that simple, because in this case the total random process at the system input, $X(t) + N(t)$, is not white. If it were white, its autocorrelation function would be a δ function and, hence, all future values of the input would be uncorrelated with the present and past values. Thus, a system that could not respond to future inputs (that is, a causal system) would not be ignoring any information that might lead to a better estimate of the signal. It appears, therefore, that the first step in obtaining a causal system should be to transform the spectral density of signal plus noise into white noise. Hence, a prewhitening filter is needed.

From (8-38), it is apparent that if one had a filter with a transfer function of

$$H_1(s) = \frac{1}{F_i(s)} \qquad \text{(8-41)}$$

then the output of this filter would be white, because

$$[S_X(s) + S_N(s)]H_1(s)H_1(-s) = \frac{S_X(s) + S_N(s)}{F_i(s)F_i(-s)} = 1$$

Furthermore, $H_1(s)$ would be causal because $F_i(s)$ by definition has only left half plane poles and zeros. Thus, $H_1(s)$ is the prewhitening filter for the input signal plus noise.

Next, look once more at the factor in (8-39) that was set equal to zero; that is,

$$F_i(s)H(s) - \frac{S_X(s)}{F_i(-s)}$$

The source of the right half plane poles is the second term of this factor, but that term can be broken up (by means of a partial fraction expansion) into the *sum* of one term having only left half plane poles and one having only right half plane poles. Thus, write this factor as

$$F_i(s)H(s) - \frac{S_X(s)}{F_i(-s)} = F_i(s)H(s) - \left[\frac{S_X(s)}{F_i(-s)}\right]_L - \left[\frac{S_X(s)}{F_i(-s)}\right]_R \qquad \text{(8-42)}$$

where the sub L implies left half poles only and the sub R implies right half plane poles only.   It is now clear that it is not possible to make this entire factor zero with a causal $H(s)$, and that the smallest value that it can have is obtained by making the difference between the first two terms of the right side of (8-42) equal to zero.   That is, let

$$F_i(s)H(s) - \left[\frac{S_X(s)}{F_i(-s)}\right]_{\mathrm{L}} = 0$$

or

$$H(s) = \frac{1}{F_i(s)}\left[\frac{S_X(s)}{F_i(-s)}\right]_{\mathrm{L}} \tag{8-43}$$

Note that the first factor of (8-43) is $H_1(s)$, the prewhitening filter. Thus, the elimination of the *noncausal* parts of the second factor represents the best that can be done in minimizing the total mean-square error.

The optimum filter, which minimizes total mean-square error, is often referred to as the *Wiener filter*.   It can be considered as a cascade of two parts, as shown in Figure 8-12.   The first part is the prewhitening filter

$$F_I(s)F_i(-s) = S_X(s) + S_N(s)$$

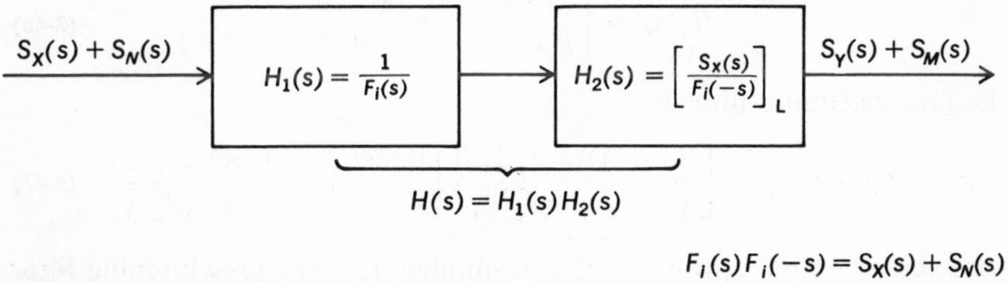

Figure 8-12   The optimum Wiener filter.

$H_1(s)$, while the second part, $H_2(s)$, does the actual filtering.   Often $H_1(s)$ and $H_2(s)$ have common factors that cancel to yield an $H(s)$ that is simpler than might be expected (and easier to build than either factor).

As an example of the Wiener filter, consider a signal having a spectral density of

$$S_X(s) = \frac{-1}{s^2 - 1}$$

and noise with a spectral density of

$$S_N(s) = \frac{-1}{s^2 - 4}$$

Thus,

$$F_i(s)F_i(-s) = S_X(s) + S_N(s) = \frac{-1}{s^2 - 1} + \frac{-1}{s^2 - 4} = \frac{-(2s^2 - 5)}{(s^2 - 1)(s^2 - 4)}$$

$P_{rr}^{+}(\omega)\, \phi\, \bar{rr}(\omega) =$

from which it follows that

$$F_i(s) = \frac{\sqrt{2}\,(s + \sqrt{2.5})}{(s + 1)(s + 2)} \tag{8-44}$$

Therefore, the prewhitening filter is

$$H_1(s) = \frac{1}{F_i(s)} = \frac{(s + 1)(s + 2)}{\sqrt{2}\,(s + \sqrt{2.5})} \tag{8-45}$$

The second filter section is obtained readily from

$$\frac{S_X(s)}{F_i(-s)} = \frac{-1}{s^2 - 1} \cdot \frac{(-s + 1)(-s + 2)}{\sqrt{2}\,(-s + \sqrt{2.5})} = \frac{s - 2}{\sqrt{2}\,(s + 1)(s - \sqrt{2.5})}$$

which may be broken up by means of a partial fraction expansion into

$$\frac{S_X(s)}{F_i(-s)} = \frac{0.822}{s + 1} - \frac{0.115}{s - \sqrt{2.5}}$$

Hence,

$$H_2(s) = \left[\frac{S_X(s)}{F_i(-s)}\right]_L = \frac{0.822}{s + 1} \tag{8-46}$$

The final optimum filter is

$$H(s) = H_1(s)H_2(s) = \left[\frac{(s + 1)(s + 2)}{\sqrt{2}\,(s + \sqrt{2.5})}\right]\left[\frac{0.822}{s + 1}\right] = \frac{0.582(s + 2)}{s + \sqrt{2.5}} \tag{8-47}$$

Note that the final optimum filter is simpler than the prewhitening filter and can be built as an RC circuit.

The remaining problem is that of evaluating the performance of the optimum filter; that is, to determine the actual value of the minimum mean-square error. This problem is greatly simplified by recognizing that in an optimum system of this sort, the error that remains must be uncorrelated with the actual output of the system. If this were not true it would be possible to perform some further linear operation on the output and obtain a still smaller error. Thus, the minimum mean-square error is simply the difference between the mean-square value of the input *signal* component and the mean-square value of the *total* filter output. That is,

$$(\overline{E^2} + \overline{M^2})_{\min}$$

$$= \frac{1}{2\pi j}\int_{-j\infty}^{j\infty} S_X(s)\,ds - \frac{1}{2\pi j}\int_{-j\infty}^{j\infty} [S_X(s) + S_N(s)]H(s)H(-s)\,ds \tag{8-48}$$

when $H(s)$ is as given by (8-43).

The above result can be used to evaluate the minimum mean-square error that is achieved by the Wiener filter described by (8-47). The first integral in (8-48) is evaluated easily by using either Table 6-1 or by summing the residues. Thus,

$$\frac{1}{2\pi j} \int_{-j\infty}^{j\infty} S_X(s) \, ds = \frac{1}{2\pi j} \int_{-j\infty}^{j\infty} \frac{-1}{s^2 - 1} \, ds = 0.5$$

The second integral is similarly evaluated as

$$\frac{1}{2\pi j} \int_{-j\infty}^{j\infty} [S_X(s) + S_N(s)] H(s) H(-s) \, ds$$

$$= \frac{1}{2\pi j} \int_{-j\infty}^{j\infty} \frac{-(2s^2 - 5)}{(s^2 - 1)(s^2 - 4)} \cdot \frac{(0.582)^2(s^2 - 4)}{(s^2 - 2.5)} \, ds$$

$$= \frac{1}{2\pi j} \int_{-j\infty}^{j\infty} \frac{-2(0.582)^2}{(s^2 - 1)} \, ds = 0.339$$

The minimum mean-square error now becomes

$$(\overline{E^2} + \overline{M^2})_{\min} = 0.5 - 0.339 = 0.161$$

It is of interest to compare this value with the mean-square error that would result if no filter were used. With no filtering there would be no signal error and the total mean-square error would be the mean-square value of the noise. Thus

$$(\overline{E^2} + \overline{M^2}) = \overline{N^2} = \frac{1}{2\pi j} \int_{-j\infty}^{j\infty} \frac{-1}{s^2 - 4} \, ds = 0.25$$

and it is seen that the use of the filter has substantially reduced the total error. This reduction would have been even more pronounced had the input noise had a wider bandwidth.

Exercise 8-6

In the example worked out in this section, suppose the signal spectral density is the same but the noise is white with a spectral density of

$$S_N(s) = 1$$

(a) Find the transfer function of the Wiener filter that minimizes the total mean-square error and (b) evaluate this minimum mean-square error.

Answer: $0.413/(s + \sqrt{2})$, 0.0857

■ PROBLEMS

8-1 Two possible criteria for optimum systems are:
A. Maximum output signal-to-noise ratio at a specified time.
B. Minimum mean-square error between the total output and the desired signal output.

For each of the following practical situations, state which criterion would be most appropriate:

a. Recovery of speech signals in an AM broadcasting system.
b. Detection of binary pulses in a digital communication system.
c. An automatic control system subject to random disturbances.
d. An aircraft flight-control system.
e. A particle detector for measuring nuclear radiation.
f. A pulse radar system.
g. A police speed radar system.
h. A radio astronomy telescope.
i. A seismic recording system.
j. A passive sonar system for detecting underwater sounds.

8-2 The impulse response, $h(t)$, in the system shown is causal and the input noise $N(t)$ is zero-mean, Gaussian and white. Prove that the out-

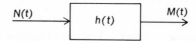

put $M(t)$ is *independent* of $N(t + \tau)$ for all $\tau > 0$ (that is, future values of the input) but is *not* independent of $N(t + \tau)$ for $\tau \le 0$ (that is, past and present values of the input).

8-3 A sinusoidal signal having a frequency of ω_0 radians per second is combined with white noise having a spectral density of N_0. It is desired to use a single-section, RC, low-pass filter to maximize the ratio of *average* signal power to average noise power. What should be the time constant of this filter?

8-4 A random signal has sample functions of the form

$$X(t) = Ytu(t)$$

where Y is a random variable uniformly distributed in the range from -3 to $+3$. This signal is added to white noise having a spectral density of 1.5 V²/Hz. A single-section, RC, low-pass filter is to be used to minimize the total mean-square error at any time after the initial transient has essentially died out. (Note that the *signal error* approaches a constant depending on Y and the time constant.) Find the time constant of the optimum filter.

8-5 A signal, $s(t)$, having the form shown is combined with white noise having a spectral density of 0.1 V²/H₂.

a. Find the impulse response of the filter that will maximize the signal-to-noise ratio at $t_0 = 1$.

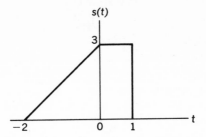

b. Find the value of this maximum signal-to-noise ratio.

c. Repeat (a) and (b) for $t_0 = 0$.

8-6   A signal has the form

$$s(t) = t\epsilon^{-t}u(t)$$

and is combined with white noise having a spectral density of 0.005 V²/Hz.

a. What is the largest output signal-to-noise ratio that could be achieved with any matched filter?

b. For what observation time, t_0, should a matched filter be constructed in order to achieve an output signal-to-noise ratio that is 0.9 of that determined in (a)?

8-7   A random signal has a spectral density of the form

$$S_X(s) = \frac{-2s^2}{s^4 - 13s^2 + 36}$$

It is combined with white noise having a spectral density of unity.   Find the transfer function of the Wiener filter that will minimize the mean-square error between the filter output and the input signal.

■ REFERENCES

See References for Chapter 1, particularly Davenport and Root, Lanning and Battin, Papoulis, and Thomas.

Appendixes

Appendixes

A

Mathematical Tables

Table A-1 Trigonometric Identities

$\sin (A \pm B) = \sin A \cos B \pm \cos A \sin B$

$\cos (A \pm B) = \cos A \cos B \mp \sin A \sin B$

$\cos A \cos B = \dfrac{1}{2} \left[\cos (A + B) + \cos (A - B) \right]$

$\sin A \sin B = \dfrac{1}{2} \left[\cos (A - B) - \cos (A + B) \right]$

$\sin A \cos B = \dfrac{1}{2} \left[\sin (A + B) + \sin (A - B) \right]$

$\sin A + \sin B = 2 \sin \dfrac{1}{2} (A + B) \cos \dfrac{1}{2} (A - B)$

$\sin A - \sin B = 2 \sin \dfrac{1}{2} (A - B) \cos \dfrac{1}{2} (A + B)$

$\cos A + \cos B = 2 \cos \dfrac{1}{2} (A + B) \cos \dfrac{1}{2} (A - B)$

$\cos A - \cos B = -2 \sin \dfrac{1}{2} (A + B) \sin \dfrac{1}{2} (A - B)$

$\sin 2A = 2 \sin A \cos A$

$\cos 2A = 2 \cos^2 A - 1 = 1 - 2 \sin^2 A = \cos^2 A - \sin^2 A$

$\sin \dfrac{1}{2} A = \sqrt{\dfrac{1}{2} (1 - \cos A)}$

$\cos \dfrac{1}{2} A = \sqrt{\dfrac{1}{2} (1 + \cos A)}$

$\sin^2 A = \dfrac{1}{2} (1 - \cos 2A)$

Table A-1   (continued)

$$\cos^2 A = \frac{1}{2}(1 + \cos 2A)$$

$$\sin x = \frac{\epsilon^{jx} - \epsilon^{-jx}}{2j} \quad \text{and} \quad \cos x = \frac{\epsilon^{jx} + \epsilon^{-jx}}{2}$$

$$\epsilon^{jx} = \cos x + j \sin x$$

$$A \cos(\omega t + \phi_1) + B \cos(\omega t + \phi_2) = C \cos(\omega t + \phi_3)$$
$$\text{where } C = \sqrt{A^2 + B^2 - 2AB \cos(\phi_2 - \phi_1)}$$
$$\phi_3 = \tan^{-1}\left[\frac{A \sin \phi_1 + B \sin \phi_2}{A \cos \phi_1 + B \cos \phi_2}\right]$$

$$\sin(\omega t + \phi) = \cos(\omega t + \phi - 90°)$$

Table A-2   Indefinite Integrals

$$\int \sin ax \, dx = -\frac{1}{a}\cos ax \qquad \int \cos ax \, dx = \frac{1}{a}\sin ax$$

$$\int \sin^2 ax \, dx = \frac{x}{2} - \frac{\sin 2ax}{4a}$$

$$\int x \sin ax \, dx = \frac{1}{a^2}(\sin ax - ax \cos ax)$$

$$\int x^2 \sin ax \, dx = \frac{1}{a^3}(2ax \sin ax + 2 \cos ax - a^2x^2 \cos ax)$$

$$\int \cos^2 ax \, dx = \frac{x^2}{2} + \frac{\sin 2ax}{4a}$$

$$\int x \cos ax \, dx = \frac{1}{a^2}(\cos ax + ax \sin ax)$$

$$\int x^2 \cos ax \, dx = \frac{1}{a^3}(2ax \cos ax - 2 \sin ax + a^2x^2 \sin ax)$$

$$\int \sin ax \sin bx \, dx = \frac{\sin(a-b)x}{2(a-b)} - \frac{\sin(a+b)}{2(a+b)}$$

$$\int \sin ax \cos bx \, dx = -\left[\frac{\cos(a-b)x}{2(a-b)} + \frac{\cos(a+b)x}{2(a+b)}\right] \qquad a^2 \neq b^2$$

$$\int \cos ax \cos bx \, dx = \frac{\sin(a-b)x}{2(a-b)} + \frac{\sin(a+b)x}{2(a+b)}$$

$$\int \epsilon^{ax} \, dx = \frac{1}{a}\epsilon^{ax}$$

Table A-2   (continued)

$$\int x\epsilon^{ax}\, dx = \frac{\epsilon^{ax}}{a^2}\,(ax - 1)$$

$$\int x^2\epsilon^{ax}\, dx = \frac{\epsilon^{ax}}{a^3}\,(a^2x^2 - 2ax + 2)$$

$$\int \epsilon^{ax} \sin bx\, dx = \frac{\epsilon^{ax}}{a^2 + b^2}\,(a \sin bx - b \cos bx)$$

$$\int \epsilon^{ax} \cos bx\, dx = \frac{\epsilon^{ax}}{a^2 + b^2}\,(a \cos bx + b \sin bx)$$

Table A-3   Definite Integrals

$$\int_0^\infty x^n\epsilon^{-ax}\, dx = \frac{n!}{a^{n+1}} = \frac{\Gamma(n + 1)}{a^{n+1}}$$

$$\text{where } \Gamma(u) = \int_0^\infty z^{u-1}\epsilon^{-z}\, dz \text{ (Gamma function)}$$

$$\int_0^\infty \epsilon^{-r^2x^2}\, dx = \frac{\sqrt{\pi}}{2r}$$

$$\int_0^\infty x\epsilon^{-r^2x^2}\, dx = \frac{1}{2r^2}$$

$$\int_0^\infty x^2\epsilon^{-r^2x^2}\, dx = \frac{\sqrt{\pi}}{4r^3}$$

$$\int_0^\infty x^n\epsilon^{-r^2x^2}\, dx = \frac{\Gamma[(n + 1)/2]}{2r^{n+1}}$$

$$\int_0^\infty \frac{\sin ax}{x}\, dx = \frac{\pi}{2},\, 0,\, -\frac{\pi}{2} \quad \text{for} \quad a > 0,\, a = 0,\, a < 0$$

$$\int_0^\infty \frac{\sin^2 x}{x}\, dx = \frac{\pi}{2}$$

$$\int_0^\infty \frac{\sin^2 ax}{x^2}\, dx = |a|\frac{\pi}{2}$$

$$\int_0^\pi \sin^2 mx\, dx = \int_0^\pi \sin^2 x\, dx = \int_0^\pi \cos^2 mx\, dx = \int_0^\pi \cos^2 x\, dx = \frac{\pi}{2},$$
$$m \text{ an integer}$$

$$\int_0^\pi \sin mx \sin nx\, dx = \int_0^\pi \cos mx \cos nx\, dx = 0 \quad m \neq n$$
$$m, n \text{ integers}$$

$$\int_0^\pi \sin mx \cos nx\, dx = \begin{cases} \dfrac{2m}{m^2 - n^2} & \text{if } m + n \text{ odd} \\ 0 & \text{if } m + n \text{ even} \end{cases}$$

Table A-4   Fourier Transforms

DESCRIPTION	$f(t)$	$F(\omega)$
Definition	$f(t) = \dfrac{1}{2\pi}\displaystyle\int_{-\infty}^{\infty} F(\omega)\epsilon^{j\omega t}\,dt$	$F(\omega) = \displaystyle\int_{-\infty}^{\infty} f(t)\epsilon^{-j\omega t}\,dt$
Reversal	$f(-t)$	$F(-\omega)$
Symmetry	$F(t)$	$2\pi f(-\omega)$
Scaling	$f(at)$	$\dfrac{1}{|a|} F\left(\dfrac{\omega}{a}\right)$
Delay	$f(t - t_0)$	$\epsilon^{-j\omega t_0}F(\omega)$
Complex conjugate	$f^*(t)$	$F^*(-\omega)$
Time differentiation	$\dfrac{d^n f(t)}{dt^n}$	$(j\omega)^n F(\omega)$
Frequency differentiation	$t^n f(t)$	$(j)^n \dfrac{d^n F(\omega)}{d\omega^n}$
Time integration	$\displaystyle\int_{-\infty}^{t} f(t)\,dt$	$\dfrac{1}{j\omega} F(\omega) + \pi F(0)\delta(\omega)$
Time convolution	$\displaystyle\int_{-\infty}^{\infty} f_1(\lambda)f_2(t - \lambda)\,d\lambda$	$F_1(\omega)F_2(\omega)$
Frequency convolution	$f_1(t)f_2(t)$	$\dfrac{1}{2\pi}\displaystyle\int_{-\infty}^{\infty} F_1(\xi)F_2(\omega - \xi)\,d\xi$
Parseval's theorem	$\displaystyle\int_{-\infty}^{\infty} f_1(t)f_2(t)\,dt = \dfrac{1}{2\pi}\displaystyle\int_{-\infty}^{\infty} F_1(\omega)F_2(-\omega)\,d\omega$	
Modulation	$\epsilon^{j\omega_0 t}f(t)$	$F(\omega - \omega_0)$
Unit impulse	$\delta(t)$	1
Unit step	$u(t)$	$\pi\delta(\omega) + \dfrac{1}{j\omega}$
Signum function	$\operatorname{sgn} t$	$\dfrac{2}{j\omega}$
Sine	$\sin \omega_0 t$	$-j\pi[\delta(\omega - \omega_0) - \delta(\omega + \omega_0)]$
Cosine	$\cos \omega_0 t$	$\pi[\delta(\omega - \omega_0) + \delta(\omega + \omega_0)]$
Rectangular pulse		$T\dfrac{\sin (\omega T/2)}{\omega T/2}$
Triangular pulse		$\dfrac{T}{2}\left(\dfrac{\sin (\omega T/4)}{\omega T/4}\right)^2$
Gaussian pulse	$\epsilon^{-\alpha^2 t^2}$	$\dfrac{\sqrt{\pi}}{\alpha} \epsilon^{-(\omega^2/4\alpha^2)}$
Fourier series	$\displaystyle\sum_{n=-\infty}^{\infty} \alpha_n \epsilon^{j(2\pi n t/T)}$	$2\pi \displaystyle\sum_{n=-\infty}^{\infty} \alpha_n \delta\left(\omega - \dfrac{2\pi n}{T}\right)$
Impulse train	$\displaystyle\sum_{n=-\infty}^{\infty} \delta(t - nT)$	$\dfrac{2\pi}{T} \displaystyle\sum_{n=-\infty}^{\infty} \delta\left(\omega - \dfrac{2\pi n}{T}\right)$
Constant	K	$2\pi K\delta(\omega)$

Table A-5   One-sided Laplace Transforms

DESCRIPTION	$f(t)$	$F(s)$
Definition	$f(t) = \dfrac{1}{2\pi j} \displaystyle\int_{c-j\infty}^{c+j\infty} F(s)\, \epsilon^{st}\, ds$	$F(s) = \displaystyle\int_{0}^{\infty} f(t)\epsilon^{-st}\, dt$
Derivative	$f'(t) = \dfrac{df(t)}{dt}$	$sF(s) - f(0)$
2nd derivative	$f''(t) = \dfrac{d^2 f(t)}{dt^2}$	$s^2 F(s) - sf(0) - f'(0)$
Integral	$\displaystyle\int_{0}^{t} f(\xi)\, d\xi$	$\dfrac{1}{s} F(s)$
t multiplication	$tf(t)$	$-\dfrac{dF(s)}{ds}$
Division by t	$\dfrac{1}{t} f(t)$	$\displaystyle\int_{s}^{\infty} F(\xi)\, d\xi$
Delay	$f(t - t_0)u(t - t_0)$	$\epsilon^{-st_0}F(s)$
Exponential decay	$\epsilon^{-at}f(t)$	$F(s + a)$
Scale change	$f(at)\quad a > 0$	$\dfrac{1}{a} F\left(\dfrac{s}{a}\right)$
Convolution	$\displaystyle\int_{0}^{t} f_1(\lambda)f_2(t - \lambda)\, d\lambda$	$F_1(s)F_2(s)$
Initial value	$f(0^+)$	$\displaystyle\lim_{s \to \infty} sF(s)$
Final value	$f(\infty)$	$\displaystyle\lim_{s \to 0} sF(s)$ [$F(s)$-Left-half plane poles only]
Impulse	$\delta(t)$	1
Step	$u(t)$	$\dfrac{1}{s}$
Ramp	$tu(t)$	$\dfrac{1}{s^2}$
nth order ramp	$t^n u(t)$	$\dfrac{n!}{s^{n+1}}$
Exponential	$\epsilon^{-\alpha t}u(t)$	$\dfrac{1}{s + \alpha}$
Damped ramp	$t\epsilon^{-\alpha t}u(t)$	$\dfrac{1}{(s + \alpha)^2}$
Sine wave	$\sin(\beta t)u(t)$	$\dfrac{\beta}{s^2 + \beta^2}$
Cosine wave	$\cos(\beta t)u(t)$	$\dfrac{s}{s^2 + \beta^2}$
Damped sine	$\epsilon^{-\alpha t}\sin(\beta t)u(t)$	$\dfrac{\beta}{(s + \alpha)^2 + \beta^2}$
Damped cosine	$\epsilon^{-\alpha t}\cos(\beta t)u(t)$	$\dfrac{s + \alpha}{(s + \alpha)^2 + \beta^2}$

APPENDIX

B

Frequently Encountered Probability Distributions

There are a number of probability distributions that occur quite frequently in the application of probability theory to practical problems. Mathematical expressions for the most common of these distributions are collected here along with their most important parameters.

The following notation is used throughout:

$\Pr(x)$—probability of the event x occurring.

$p_X(x)$—probability density function of the random variable X at the point x.

$\bar{X} = E\{X\}$—mean of the random variable X.

$\sigma_X^2 = E\{(X - \bar{X})^2\}$—variance of the random variable X.

$\phi(u) = \displaystyle\int_{-\infty}^{\infty} p_X(x)\epsilon^{jux}\,dx$—characteristic function of the random variable X.

Discrete Probability Functions

Bernoulli (Special case of Binomial)

$$\Pr(x) = \begin{cases} p & x = 1 \\ q = 1 - p & x = 0 \\ 0 & \text{otherwise} \end{cases} \qquad 0 < p < 1$$

$p_X(x) = p\delta(x - 1) + q\delta(x)$

$\bar{X} = p$

$\sigma_X^2 = pq$

$\phi(u) = 1 - p + p\epsilon^{ju}$

Binomial

$$\Pr(x) = \begin{cases} \dbinom{n}{x} p^x q^{n-x} & x = 0, 1, 2, \ldots, n \\ 0 & \text{otherwise} \end{cases}$$

$$0 < p < 1 \qquad q = 1 - p \qquad n = 1, 2, \ldots$$

Binomial (continued)

$$p_X(x) = \sum_{k=0}^{n} \binom{n}{k} p^k q^{n-k} \delta(x - k)$$

$$\bar{X} = np$$

$$\sigma_X{}^2 = npq$$

$$\phi(u) = [1 - p + p\epsilon^{ju}]^n$$

Pascal

$$\Pr(x) = \begin{cases} \binom{x-1}{n-1} p^n q^{x-n} & x = n, \, n+1, \, \ldots \\ 0 & \text{otherwise} \end{cases}$$

$$0 < p < 1 \qquad q = 1 - p \qquad n = 1, 2, 3, \, \ldots$$

$$\bar{X} = np^{-1}$$

$$\sigma_X{}^2 = nqp^{-2}$$

$$\phi(u) = p^n \epsilon^{jnu}[1 - q\epsilon^{ju}]^{-n}$$

Poisson

$$\Pr(x) = \frac{a^x \epsilon^{-a}}{x!} \qquad x = 0, 1, 2, \, \ldots$$

$$a > 0$$

$$\bar{X} = a$$

$$\sigma_X{}^2 = a$$

$$\phi(u) = \epsilon^{a(\epsilon^{ju}-1)}$$

Continuous Distributions

Beta

$$p_X(x) = \begin{cases} \dfrac{(a+b-1)!}{(a-1)!(b-1)!} x^{a-1}(1-x)^{b-1} & 0 < x < 1 \\ 0 & \text{otherwise} \end{cases}$$

$$a > 0 \qquad b > 0$$

$$\bar{X} = \frac{a}{a+b}$$

$$\sigma_X{}^2 = \frac{ab}{(a+b)^2(a+b+1)}$$

Cauchy

$$p_X(x) = \frac{1}{\pi} \frac{a}{a^2 + (x-b)^2} \qquad -\infty < x < \infty$$

$$a > 0 \qquad -\infty < b < \infty$$

Mean and variance not defined

$$\phi(u) = \epsilon^{jbu - a|u|}$$

Chi-square

$$p_X(x) = \begin{cases} \left[\left(\dfrac{n}{2} - 1\right)!\right]^{-1} 2^{-n/2} x^{(n/2)-1} \epsilon^{-x/2} & x > 0 \\ 0 & \text{otherwise} \end{cases}$$

$$n = 1, 2, \ldots$$
$$\bar{X} = n$$
$$\sigma_X{}^2 = 2n$$
$$\phi(u) = (1 - 2ju)^{-n/2}$$

Erlang

$$p_X(x) = \begin{cases} \dfrac{a^n x^{n-1} \epsilon^{-ax}}{(n-1)!} & x > 0 \\ 0 & \text{otherwise} \end{cases}$$

$$a > 0 \qquad n = 1, 2, \ldots$$
$$\bar{X} = na^{-1}$$
$$\sigma_X{}^2 = na^{-2}$$
$$\phi(u) = a^n (a - ju)^{-n}$$

Exponential

$$p_X(x) = \begin{cases} a\epsilon^{-ax} & x > 0 \\ 0 & \text{otherwise} \end{cases}$$

$$a > 0$$
$$\bar{X} = a^{-1}$$
$$\sigma_X{}^2 = a^{-2}$$
$$\phi(u) = a(a - ju)^{-1}$$

Gamma

$$p_X(x) = \begin{cases} \dfrac{x^a \epsilon^{-x/b}}{a! b^{a+1}} & x > 0 \\ 0 & \text{otherwise} \end{cases}$$

$$a > -1 \qquad b > 0$$
$$\bar{X} = (a + 1)/b$$
$$\sigma_X{}^2 = (a + 1)b^2$$
$$\phi(u) = (1 - jbu)^{-a-1}$$

Laplace

$$p_X(x) = \frac{a}{2} \epsilon^{-a|x-b|} \qquad -\infty < x < \infty$$

$$a > 0 \qquad -\infty < b < \infty$$
$$\bar{X} = b$$
$$\sigma_X{}^2 = 2a^{-2}$$
$$\phi(u) = a^2 \epsilon^{jbu}(a^2 + u^2)^{-1}$$

Log-normal

$$p_X(x) = \begin{cases} \dfrac{\exp\{-[\ln(x-a)-b]^2/2\sigma^2\}}{\sqrt{2\pi}\,\sigma(x-a)} & x \geq a \\ 0 & \text{otherwise} \end{cases}$$

$$\sigma > 0 \qquad -\infty < a < \infty \qquad -\infty < b < \infty$$

$$\bar{X} = a + \epsilon^{b+.5\sigma^2}$$

$$\sigma_X{}^2 = \epsilon^{2b+\sigma^2}(\epsilon^{\sigma^2} - 1)$$

Maxwell

$$p_X(x) = \begin{cases} \sqrt{\dfrac{2}{\pi}}\,a^3x^2\epsilon^{-a^2x^2/2} & x > 0 \\ 0 & \text{otherwise} \end{cases}$$

$$a > 0$$

$$\bar{X} = \sqrt{8/\pi}\,a^{-1}$$

$$\sigma_X{}^2 = \left(3 - \dfrac{8}{\pi}\right)a^{-2}$$

Normal

$$p_X(x) = \frac{1}{\sqrt{2\pi}\,\sigma_X}\,\epsilon^{-(x-\bar{X})^2/2\sigma_X{}^2} \qquad -\infty < x < \infty$$

$$\sigma_X > 0 \qquad -\infty < \bar{X} < \infty$$

$$\phi(u) = \epsilon^{ju\bar{X} - (u^2\sigma_X{}^2/2)}$$

Normal-bivariate

$$p_{X,Y}(x,y) = \frac{1}{2\pi\sigma_X\sigma_Y\sqrt{1-\rho^2}}\exp\left\{\frac{-1}{2(1-\rho^2)}\left[\left(\frac{x-\bar{X}}{\sigma_X}\right)^2 + \left(\frac{y-\bar{Y}}{\sigma_Y}\right)^2\right.\right.$$

$$\left.\left. - \frac{2\rho}{\sigma_X\sigma_Y}(x-\bar{X})(y-\bar{Y})\right]\right\}$$

$$-\infty < x < \infty \qquad -\infty < y < \infty \qquad \sigma_X > 0 \qquad \sigma_Y > 0$$

$$-1 < \rho < 1$$

$$\phi(u,v) = \exp\left[ju\bar{X} + jv\bar{Y} - \tfrac{1}{2}(u^2\sigma_X{}^2 + 2\rho uv\sigma_X\sigma_Y + v^2\sigma_Y{}^2)\right]$$

Rayleigh

$$p_X(x) = \begin{cases} \dfrac{x}{a^2}\,\epsilon^{-x^2/2a^2} & x > 0 \\ 0 & \text{otherwise} \end{cases}$$

$$\bar{X} = a\sqrt{\frac{\pi}{2}}$$

$$\sigma_X{}^2 = \left(2 - \frac{\pi}{2}\right)a^2$$

Uniform

$$p_X(x) = \begin{cases} \dfrac{1}{b-a} & a < x < b \\ 0 & \text{otherwise} \end{cases}$$

$$-\infty < a < b < \infty$$

$$\bar{X} = \frac{a+b}{2}$$

$$\sigma_X{}^2 = \frac{(b-a)^2}{12}$$

$$\phi(u) = \frac{\epsilon^{jub} - \epsilon^{jua}}{ju(b-a)}$$

Weibull

$$p_X(x) = \begin{cases} abx^{b-1}\epsilon^{-ax^b} & x > 0 \\ 0 & \text{otherwise} \end{cases}$$

$$a > 0 \qquad b > 0$$

$$\bar{X} = \left(\frac{1}{a}\right)^{1/b} \Gamma(1 + b^{-1})$$

$$\sigma_X{}^2 = \left(\frac{1}{a}\right)^{2/b} \{\Gamma(1 + 2b^{-1}) - [\Gamma(1 + b^{-1})]^2\}$$

Binomial Coefficients

$$\binom{n}{m} = \frac{n!}{(n-m)!\,m!}$$

n	$\binom{n}{0}$	$\binom{n}{1}$	$\binom{n}{2}$	$\binom{n}{3}$	$\binom{n}{4}$	$\binom{n}{5}$	$\binom{n}{6}$	$\binom{n}{7}$	$\binom{n}{8}$	$\binom{n}{9}$
0	1									
1	1	1								
2	1	2	1							
3	1	3	3	1						
4	1	4	6	4	1					
5	1	5	10	10	5	1				
6	1	6	15	20	15	6	1			
7	1	7	21	35	35	21	7	1		
8	1	8	28	56	70	56	28	8	1	
9	1	9	36	84	126	126	84	36	9	1
10	1	10	45	120	210	252	210	120	45	10
11	1	11	55	165	330	462	462	330	165	55
12	1	12	66	220	495	792	924	792	495	220
13	1	13	78	286	715	1287	1716	1716	1287	715
14	1	14	91	364	1001	2002	3003	3432	3003	2002
15	1	15	105	455	1365	3003	5005	6435	6435	5005
16	1	16	120	560	1820	4368	8008	11440	12870	11440
17	1	17	136	680	2380	6188	12376	19448	24310	24310
18	1	18	153	816	3060	8568	18564	31824	43758	48620
19	1	19	171	969	3876	11628	27132	50388	75582	92378

$$\binom{n}{m} + \binom{n}{m+1} = \binom{n+1}{m+1}$$

Useful Relationships

$$\binom{n}{n-m} = \binom{n}{m}$$

D

Normal Probability Distribution Function

$$\Phi(x) = \frac{1}{\sqrt{2\pi}} \int_{-\infty}^{x} \epsilon^{-t^2/2} \, dt; \quad \Phi(-x) = 1 - \Phi(x)$$

x	.00	.01	.02	.03	.04	.05	.06	.07	.08	.09
0.0	.5000	.5040	.5080	.5120	.5160	.5199	.5239	.5279	.5319	.5359
0.1	.5398	.5438	.5478	.5517	.5557	.5596	.5636	.5675	.5714	.5753
0.2	.5793	.5832	.5871	.5910	.5948	.5987	.6026	.6064	.6103	.6141
0.3	.6179	.6217	.6255	.6293	.6331	.6368	.6406	.6443	.6480	.6517
0.4	.6554	.6591	.6628	.6664	.6700	.6736	.6772	.6808	.6844	.6879
0.5	.6915	.6950	.6985	.7019	.7054	.7088	.7123	.7157	.7190	.7224
0.6	.7257	.7291	.7324	.7357	.7389	.7422	.7454	.7486	.7517	.7549
0.7	.7580	.7611	.7642	.7673	.7704	.7734	.7764	.7794	.7823	.7852
0.8	.7881	.7910	.7939	.7967	.7995	.8023	.8051	.8078	.8106	.8133
0.9	.8159	.8186	.8212	.8238	.8264	.8289	.8315	.8340	.8365	.8389
1.0	.8413	.8438	.8461	.8485	.8508	.8531	.8554	.8577	.8599	.8621
1.1	.8643	.8665	.8686	.8708	.8729	.8749	.8770	.8790	.8810	.8830
1.2	.8849	.8869	.8888	.8907	.8925	.8944	.8962	.8980	.8997	.9015
1.3	.9032	.9049	.9066	.9082	.9099	.9115	.9131	.9147	.9162	.9177
1.4	.9192	.9207	.9222	.9236	.9251	.9265	.9279	.9292	.9306	.9319
1.5	.9332	.9345	.9357	.9370	.9382	.9394	.9406	.9418	.9429	.9441
1.6	.9452	.9463	.9474	.9484	.9495	.9505	.9515	.9525	.9535	.9545
1.7	.9554	.9564	.9573	.9582	.9591	.9599	.9608	.9616	.9625	.9633
1.8	.9641	.9649	.9656	.9664	.9671	.9678	.9686	.9693	.9699	.9706
1.9	.9713	.9719	.9726	.9732	.9738	.9744	.9750	.9756	.9761	.9767
2.0	.9772	.9778	.9783	.9788	.9793	.9798	.9803	.9808	.9812	.9817
2.1	.9821	.9826	.9830	.9834	.9838	.9842	.9846	.9850	.9854	.9857
2.2	.9861	.9864	.9868	.9871	.9875	.9878	.9881	.9884	.9887	.9890
2.3	.9893	.9896	.9898	.9901	.9904	.9906	.9909	.9911	.9913	.9916
2.4	.9918	.9920	.9922	.9925	.9927	.9929	.9931	.9932	.9934	.9936
2.5	.9938	.9940	.9941	.9943	.9945	.9946	.9948	.9949	.9951	.9952
2.6	.9953	.9955	.9956	.9957	.9959	.9960	.9961	.9962	.9963	.9964
2.7	.9965	.9966	.9967	.9968	.9969	.9970	.9971	.9972	.9973	.9974
2.8	.9974	.9975	.9976	.9977	.9977	.9978	.9979	.9979	.9980	.9981
2.9	.9981	.9982	.9982	.9983	.9984	.9984	.9985	.9985	.9986	.9986
3.0	.9987	.9987	.9987	.9988	.9988	.9989	.9989	.9989	.9990	.9990
3.1	.9990	.9991	.9991	.9991	.9992	.9992	.9992	.9992	.9993	.9993
3.2	.9993	.9993	.9994	.9994	.9994	.9994	.9994	.9995	.9995	.9995
3.3	.9995	.9995	.9996	.9996	.9996	.9996	.9996	.9996	.9996	.9997
3.4	.9997	.9997	.9997	.9997	.9997	.9997	.9997	.9997	.9998	.9998
3.5	.9998	.9998	.9998	.9998	.9998	.9998	.9998	.9998	.9998	.9998
3.6	.9998	.9999	.9999	.9999	.9999	.9999	.9999	.9999	.9999	.9999
3.7	.9999	.9999	.9999	.9999	.9999	.9999	.9999	.9999	.9999	.9999
3.8	.9999	.9999	.9999	.9999	.9999	.9999	.9999	1.0000	1.0000	1.0000

Table of Correlation Function—Spectral Density Pairs

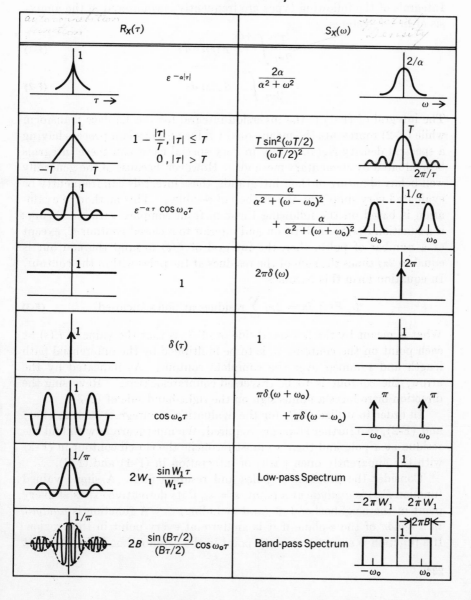

$R_X(\tau)$	$S_X(\omega)$
$\varepsilon^{-\alpha|\tau|}$	$\dfrac{2\alpha}{\alpha^2 + \omega^2}$
$1 - \dfrac{|\tau|}{T},\ |\tau| \le T$ $0,\ |\tau| > T$	$\dfrac{T \sin^2(\omega T/2)}{(\omega T/2)^2}$
$\varepsilon^{-\alpha\tau} \cos \omega_o \tau$	$\dfrac{\alpha}{\alpha^2 + (\omega - \omega_o)^2}$ $+ \dfrac{\alpha}{\alpha^2 + (\omega + \omega_o)^2}$
1	$2\pi\delta(\omega)$
$\delta(\tau)$	1
$\cos \omega_o \tau$	$\pi\delta(\omega + \omega_o)$ $+ \pi\delta(\omega - \omega_o)$
$2W_1 \dfrac{\sin W_1 \tau}{W_1 \tau}$	Low-pass Spectrum
$2B \dfrac{\sin(B\tau/2)}{(B\tau/2)} \cos \omega_o \tau$	Band-pass Spectrum

APPENDIX

F

Contour Integration

Integrals of the following types are frequently encountered in the analysis of linear systems

$$\frac{1}{2\pi j} \int_{c-j\infty}^{c+j\infty} F(s)\epsilon^{st} \, ds \tag{F-1}$$

$$\frac{1}{2\pi j} \int_{-\infty}^{\infty} S_X(s) \, ds \tag{F-2}$$

The integral of (F-1) is the inversion integral for the Laplace transform while (F-2) represents the mean-square value of a random process having a spectral density $S_X(s)$. Only in very special cases can these integrals be evaluated by elementary methods. However, because of the generally well-behaved nature of their integrands, these integrals can frequently be evaluated very simply by the method of residues. This method of evaluation is based on the following theorem from complex variable theory: if a function $F(s)$ is analytic on and interior to a closed contour C, except at a number of poles, then the integral of $F(s)$ around the contour is equal to $2\pi j$ times the sum of the residues at the poles within the contour. In equation form this becomes

$$\oint_C F(s) \, ds = 2\pi j \sum \text{residues at poles enclosed} \tag{F-3}$$

What is meant by the left-hand side of (F-3) is that the value of $F(s)$ at each point on the contour, C, is to be multiplied by the differential path length and summed over the complete contour. As indicated by the arrow, the contour is to be traversed counterclockwise. Reversing the direction introduces a minus sign on the right-hand side of (F-3).

In order to utilize (F-3) for the evaluation of integrals such as (F-1) and (F-2), two further steps are required: We must learn how to find the residue at a pole and then we must reconcile the closed contour in (F-3) with the apparently open paths of integration in (F-1) and (F-2).

Consider the problem of poles and residues first. A single valued function $F(s)$ is *analytic* at a point, $s = s_0$, if its derivative exists at every point in the neighborhood of (and including) s_o. A function is *analytic in a region* of the s-plane if it is analytic at every point in that region. If a function is analytic at every point in the neighborhood of s_0, but not at s_0 itself, then s_o is called a *singular point*. For example, the function $F(s) = 1/(s - 2)$ has a derivative $F'(s) = -1/(s - 2)^2$. It is readily seen by inspection that this function is analytic everywhere except at $s = 2$, where it has a singularity. An *isolated singular point* is a point interior to a region throughout which the function is analytic except at that point. It is evident that the above function has an isolated singularity at $s = 2$. The most frequently encountered singularity is the *pole*. If a function $F(s)$ becomes infinite at $s = s_0$ in such a manner that by multiplying $F(s)$ by a factor of the form $(s - s_0)^n$, where n is a positive integer, the singularity is removed, then $F(s)$ is said to have a pole of order n at $s = s_0$. For example, the function $1/\sin s$ has a pole at $s = 0$ and can be written as

$$F(s) = \frac{1}{\sin s} = \frac{1}{s - s^3/3! + s^5/5! - \cdots}$$

Multiplying by s [that is, the factor $(s - s_0)$] we obtain

$$\phi(s) = \frac{s}{s - s^3/3! + s^5/5! + \cdots} = \frac{1}{1 - s^2/3! + s^4/5! + \cdots}$$

which is seen to be well-behaved near $s = 0$. It may therefore be concluded that $1/\sin s$ has a simple (that is, first order) pole at $s = 0$.

It is an important property of analytic functions that they can be represented by convergent power series throughout their region of analyticity. By a simple extension of this property it is possible to represent functions in the vicinity of a singularity. Consider a function $F(s)$ having an nth order pole at $s = s_0$. Define a new function $\phi(s)$ such that

$$\phi(s) = (s - s_0)^n F(s) \tag{F-4}$$

Now $\phi(s)$ will be analytic in the region of s_0 since the singularity of $F(s)$ has been removed. Therefore, $\phi(s)$ can be expanded in a Taylor series as follows:

$$\phi(s) = A_{-n} + A_{-n+1}(s - s_0) + A_{-n+2}(s - s_0)^2$$
$$+ \cdots + A_{-1}(s - s_0)^{n-1} + \sum_{k=0}^{\infty} B_k(s - s_0)^{n+k} \tag{F-5}$$

Substituting (F-5) into (F-4) and solving for $F(s)$ gives

$$F(s) = \frac{A_{-n}}{(s - s_0)^n} + \frac{A_{-n+1}}{(s - s_0)^{n-1}} + \cdots + \frac{A_{-1}}{s - s_0} + \sum_{k=0}^{\infty} B_k(s - s_0)^k \tag{F-6}$$

This expansion is valid in the vicinity of the pole at $s = s_0$. The series converges in a region around s_0 that extends out to the nearest singularity. Equation (F-6) is called the *Laurent expansion* or *Laurent series* for $F(s)$ about the singularity at $s = s_0$. There are two distinct parts to the series: The first, called the *principal part*, consists of the terms containing $(s - s_0)$ raised to negative powers; the second, sometimes called the Taylor part, consists of terms containing $(s - s_0)$ raised to zero or positive powers. It should be noted that the second part is analytic throughout the s-plane (except at infinity) and assumes the value B at $s = s_0$. If there were no singularity in $F(s)$, only the second part of the expansion would be present and would just be the Taylor series expansion. The coefficient of $(s - s_0)^{-1}$, which is A_{-1} in (F-6), is called the *residue* of $F(s)$ of the pole at $s = s_0$.

Formally the coefficients of the Laurent series can be determined from the usual expression for the Taylor series expansion for the function $\phi(s)$ and the subsequent division by $(s - s_0)^n$. For most cases of engineering interest, simpler methods can be employed. Due to the uniqueness of the properties of analytic functions it follows that any series of the proper form [that is, the form given in (F-6)] must, in fact, be the Laurent series. When $F(s)$ is a ratio of two polynomials in s, a simple procedure for finding the Laurent series is as follows: Form $\phi(s) = (s - s_0)^n F(s)$; let $s - s_0 = v$ or $s = v + s_0$; expand $\phi(v + s_0)$ around $v = 0$ by dividing the denominator into the numerator; and replace v by $s - s_0$. As an example consider the following

$$F(s) = \frac{2}{s^2(s^2 - 1)}$$

Let it be required to find the Laurent series for $F(s)$ in the vicinity of $s = -1$:

$$\phi(s) = \frac{2}{s^2(s - 1)}$$

Let $s = v - 1$

$$\phi(v - 1) = \frac{2}{(v^2 - 2v + 1)(v - 2)} = \frac{2}{v^3 - 4v^2 - 3v - 2}$$

$$
\begin{array}{r}
-1 + \dfrac{3v}{2} - \dfrac{v^2}{4} \\[2mm]
-2 - 3v - 4v^2 + v^3 \overline{)\,2} \\[1mm]
\underline{2 + 3v + 4v^2 - v^3} \\[1mm]
-3v - 4v^2 + v^3 \\[1mm]
\underline{-3v - \dfrac{9v^2}{2} - 6v^3 + \dfrac{3v^4}{2}} \\[1mm]
\dfrac{v^2}{2} + 7v^3 - \dfrac{3v^4}{2} \\[1mm]
\underline{\dfrac{v^2}{2} + \dfrac{3v^3}{4} + v^4 - \dfrac{v^5}{4}}
\end{array}
$$

$$\phi(v - 1) = -1 + \frac{3}{2}v - \frac{1}{4}v^2 - \cdots$$

Replacing $v - 1$ by s gives

$$\phi(s) = -1 + \frac{3}{2}(s + 1) - \frac{1}{4}(s + 1)^2 - \cdots$$

$$F(s) = -\frac{1}{s + 1} + \frac{3}{2} - \frac{1}{4}(s + 1) - \cdots$$

The residue is seen to be -1.

A useful formula for finding the residue at an nth order pole, $s = s_0$, is as follows:[1]

$$K_{s_0} = \frac{\phi^{(n-1)}(s_0)}{(n - 1)!} \tag{F-7}$$

where $\phi(s) = (s - s_0)^n F(s)$. This formula is valid for $n = 1$ and is not restricted to rational functions.

When $F(s)$ is not a ratio of polynomials it is permissible to replace transcendental terms by series valid in the vicinity of the pole. For example

$$F(s) = \frac{\sin s}{s^2} = \frac{1}{s^2}\left[s - \frac{s^3}{3!} + \frac{s^5}{5!} - \cdots \right]$$

$$= \frac{1}{s} - \frac{s}{3!} + \frac{s^3}{5!} - \cdots$$

In this instance the residue of the pole at $s = 0$ is 1.

There is a direct connection between the Laurent series and the partial fraction expansion of a function $F(s)$. In particular, if $H_i(s)$ is the principal part of the Laurent series at the pole $s = s_i$, then the partial fraction expansion of $F(s)$ can be written as

$$F(s) = H_1(s) + H_2(s) \cdots H_k(s) + q(s)$$

where the first k terms are the principal parts of the Laurent series about the k poles and $q(s)$ is a polynomial $a_0 + a_1 s + a_2 s^2 + \cdots a_m s^m$ representing the behavior of $F(s)$ for large s. The value of m is the difference of the degree of the numerator polynomial minus the degree of the denominator polynomial. In general $q(s)$ can be determined by dividing the denominator polynomial into the numerator polynomial until the remainder is of lower order than the denominator. The remainder can then be expanded in its principal parts.

With the question of how to determine residues out of the way, the only remaining question is how to relate the closed contour of (F-3) with the open (straight line) contours of (F-1) and (F-2). This is handled quite easily by restricting consideration to integrands that approach zero rapidly enough for large values of the variable so that there will be no

---

[1] Where $\phi^{(n-1)}(s_o)$ denotes the $(n - 1)$-derivative of $\phi(s)$, with respect to s, evaluated at $s = s_o$.

contribution to the integral from distant portions of the contour.   Thus, although the specified path of integration in the s-plane may be from $s = c - j\infty$ to $c + j\infty$, the integral that will be evaluated will have a path of integration as shown in Figure F-1.   The path of integration

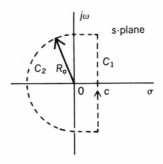

Figure F-1   Path of integration in the s-plane.

will be along the contour C_1 going from $c - jR_o$ to $c + jR_o$ and then along the contour C_2 which is a semicircle closing to the left.   The integral can then be written as

$$\oint_{C_1+C_2} F(s)\ ds = \int_{C_1} F(s)\ ds + \int_{C_2} F(s)\ ds \qquad \textbf{(F-8)}$$

If in the limit as $R_o \to \infty$ the contribution from the right-hand integral is zero, then we have

$$\int_{c-j\infty}^{c+j\infty} F(s)\ ds = \lim_{R_o \to \infty} \oint_{C_1+C_2} F(s)\ ds$$
$$= 2\pi j\ \Sigma\ \text{residues at poles enclosed} \qquad \textbf{(F-9)}$$

In any specific case the behavior of the integral over C_2 can be examined to verify that there is no contribution from this portion of the contour. However, the following two special cases will cover many of the situations in which this problem arises:

1.   Whenever $F(s)$ is a rational function having a denominator whose order exceeds the numerator by at least two, then it may be shown readily that

$$\int_{c-j\infty}^{c+j\infty} F(s)\ ds = \oint_{C_1+C_2} F(s)\ ds$$

2.   If $F_1(s)$ is analytic in the left half plane except for a finite number of poles and tends uniformly to zero as $|s| \to \infty$ with $\sigma < 0$ then for positive t the following is true (Jordan's lemma)

$$\lim_{R_o \to \infty} \int_{C_2} F_1(s)\epsilon^{st}\ ds = 0$$

From this it follows that when these conditions are met, the inversion integral of the Laplace transform can be evaluated as

$$f(t) = \frac{1}{2\pi j} \int_{c-j\infty}^{c+j\infty} F_1(s)\epsilon^{st} \, ds = \frac{1}{2\pi j} \oint_{C_1+C_2} F_1(s)\epsilon^{st} \, ds = \sum_j k_j$$

where k_j is the residue at the jth pole to the left of the abscissa of absolute convergence.

The following two examples illustrate these procedures:

Example F-1. Given that a random process has a spectral density of the following form

$$S_X(\omega) = \frac{1}{(\omega^2 + 1)(\omega^2 + 4)}$$

Find the mean-square value of the process. Converting to $S_X(s)$ gives

$$S_X(s) = \frac{1}{(-s^2 + 1)(-s^2 + 4)} = \frac{1}{(s^2 - 1)(s^2 - 4)}$$

and the mean-square value is

$$\overline{X^2} = \frac{1}{2\pi j} \int_{-j\infty}^{j\infty} \frac{ds}{(s^2 - 1)(s^2 - 4)} = k_{-1} + k_{-2}$$

From the partial fraction expansion for $S_X(s)$, the residues are found to be

$$k_{-1} = \frac{1}{(-1 - 1)(1 - 4)} = \frac{1}{6}$$

$$k_{-2} = \frac{1}{(4 - 1)(-2 - 2)} = -\frac{1}{12}$$

Therefore,

$$\overline{X^2} = \frac{1}{6} - \frac{1}{12} = \frac{1}{12}$$

Example F-2. Find the inverse Laplace transform of $F(s) = \dfrac{1}{s(s + 2)}$.

$$f(t) = \frac{1}{2\pi j} \int_{c-j\infty}^{c+j\infty} \frac{\epsilon^{st}}{s(s + 2)} \, ds = k_0 + k_{-2}$$

From (F-7)

$$k_0 = \frac{\epsilon^{st}}{s + 2}\bigg|_{s=0} = \frac{1}{2}$$

$$k_{-2} = \frac{\epsilon^{st}}{s}\bigg|_{s=-2} = \frac{\epsilon^{-2t}}{-2}$$

therefore

$$f(t) = \frac{1}{2}(1 - \epsilon^{-2t}) \qquad t > 0$$

▬REFERENCES

Churchill, R., *Operational Mathematics*, 2nd Ed., McGraw-Hill Book Co., Inc., New York, N.Y., 1958.

A thorough treatment of the mathematics relating particularly to the Laplace transform is given including an introduction to complex variable theory and contour integration.

Papoulis, A., *The Fourier Integral and its Applications*, McGraw-Hill Book Co., Inc., New York, N.Y., 1962.

In Chapter 9 and Appendix II, a readily understandable treatment of evaluation of transforms by contour integration is given.

Selected Answers

Selected Answers

Chapter 1

1-1   **a.** $\frac{1}{6}$ **b.** $\frac{1}{3}$ **c.** $\frac{1}{2}$

1-2   **a.** $\frac{1}{6}$ **b.** $\frac{13}{18}$ **c.** $\frac{1}{2}$

1-3   **a.** $\frac{13}{50}$ **b.** $\frac{3}{10}$ **c.** $\frac{1}{10}$ **d.** 0

1-4   **a.** $\frac{13}{500}$ **b.** $\frac{2}{65}$ **c.** $\frac{7}{500}$

1-5   **a.** $\dfrac{2653}{(52)^5}$ **b.** same as a.   **c.** $\frac{9}{47}$

1-7 $A + B = \{1,2,3,5,7,9\}$; $B + C = S$; $A + C = \{1,2,3,4,6,8,10\}$;
$AB = \{1,3\}$; $AC = \{2\}$; $BC = \phi$; $ABC = \phi$; $\bar{A} = \{4,5,6,7,8,9,10\}$; $\bar{B} = C$;
$\bar{C} = B$; $\bar{A}B = \{5,7,9\}$; $A\bar{B} = \{2\}$; $\overline{BC} = S$; $A - C = \{1,3\}$;
$C - A = \{4,6,8,10\}$; $A - B = \{2\}$; $(A - B) + B = \{1,2,3,5,7,9\}$;
$(A - B) + C = C$

1-10   0.0527

1-11   **a.** .787   **b.** .907

1-12   **a.** To win 3 games out of 4 is more probable.   **b.** To win at least 5 games out of 8 is more probable.

1-13   **a.** $\frac{1}{243}$ **b.** $\frac{64}{81}$

1-14   **a.** 9 channels   **b.** 4 channels

Chapter 2

2-1   **b.** $\frac{5}{12}$

2-2   **a.** .25   **b.** .75   **c.** .25

2-3   **c.** .983

2-4   **a.** $A = e$ **b.** $1 - e^{-1}$

2-5   **a.** 2   **b.** 5   **c.** 1

2-6   **a.** 2   **b.** $\frac{37}{3}$ **c.** $\frac{25}{3}$ **d.** 125

2-7   **b.** $\bar{X} = 73.13$; $\sigma = 10.25$

2-8   **a.** $p(x) = \dfrac{1}{2\sqrt{2\pi}} e^{-\frac{(x-2)^2}{8}}$ **b.** $p(0) = .1205$; $p(2) = .199$; $p(4) = .1205$

   c. $\Pr\{X > 0\} = .8413$; $\Pr\{X > 2\} = \frac{1}{2}$ **d.** .6826

2-9   **a.** 170   **b.** 57,600   **c.** If the absolute value of the current mean is increased, then the mean and variance of the power will be increased.

2-11   **a.** 10   **b.** 12.54   **c.** e^{-2}

2-12 **a.** $p(x) = \dfrac{4(x-a)}{(b-a)^2} \qquad a < x \leq \dfrac{a+b}{2}$

$\qquad\qquad = \dfrac{-4(x-b)}{(b-a)^2} \qquad \dfrac{a+b}{2} < x \leq b$

$\qquad\qquad = 0 \qquad\qquad \text{elsewhere}$

b. $\dfrac{b+a}{2}$ **c.** $(b-a)^2/24$

2-13 **a.** $p_X(x) = \dfrac{1}{\pi\sqrt{1-x^2}} \qquad -1 < x < 1$

$\qquad\qquad = 0 \qquad\qquad \text{elsewhere}$

b. 0 **c.** $\frac{1}{2}$

2-14 **a.** $p_X(x) = \dfrac{2}{\pi\sqrt{1-x^2}} \qquad 0 < x < 1$

$\qquad\qquad = 0 \qquad\qquad \text{elsewhere}$

b. $2/\pi$

2-15 **a.** $p(r|M) = 0 \qquad r \leq r_0$

$\qquad\qquad = \dfrac{p(r)}{e^{-r_0^2/2}} \quad r > r_0$

b. $r_0 + \sqrt{2\pi}\, e^{r_0^2/2}\, \Phi(-r_0)$

2-16 **b.** $p(x|M) = \dfrac{x}{4} \quad 1 < x \leq 3$

$\qquad\qquad = 0 \quad \text{elsewhere}$

2-17 **a.** $p_Y(y) = \dfrac{1}{|a|}\, p_X\left(\dfrac{y-b}{a}\right)$ **b.** $E[Y] = a\bar{X} + b;\ \sigma_Y^2 = a^2\sigma_X^2$

2-18 **a.** $10\sqrt{\pi}/2$ **b.** $10\sqrt{\pi}/2$

Chapter 3

3-1 **b.** $P(x,y) = 2xy - y^2 \quad \begin{cases} 0 < x \leq 1 \\ 0 < y \leq x \end{cases}$

$\qquad\qquad = 2y - y^2 \quad \begin{cases} x > 1 \\ 0 < y \leq 1 \end{cases}$

$\qquad\qquad = x^2 \quad \begin{cases} 0 < x \leq 1 \\ y > x \end{cases}$

$\qquad\qquad = 1 \quad \begin{cases} x > 1 \\ y > 1 \end{cases}$

$\qquad\qquad = 0 \qquad \text{elsewhere}$

c. 2 **d.** $\frac{1}{4}$ **e.** $p_X(x) = 2x \quad 0 < x \leq 1$

$\qquad\qquad = 0 \quad \text{elsewhere}$

3-2 **a.** $\frac{1}{4}$ **b.** no

3-3 **a.** 12

b. $P(x,y) = (1 - e^{-3x})(1 - e^{-4y}) \quad x > 0, y > 0$

   c. $\Pr\{0 < X \le 1, 0 < Y \le 2\} = (1 - e^{-3})(1 - e^{-8})$

   d. $p_X(x) = 3e^{-3x} \quad x > 0$
$$= 0 \qquad \text{elsewhere}$$
$$p_Y(y) = 4e^{-4y} \quad y > 0$$
$$= 0 \qquad \text{elsewhere}$$

3-4 $\quad p(x|y) = \dfrac{1}{1 - y} \quad \begin{array}{l} y < x \le 1 \\ 0 < y \le 1 \end{array}$
$$= 0 \qquad \text{elsewhere}$$

$\quad p(y|x) = \dfrac{1}{x} \quad \begin{array}{l} 0 < x \le 1 \\ 0 < y \le x \end{array}$
$$= 0 \qquad \text{elsewhere}$$

3-5 $\quad \frac{25}{6}$

3-6   **a.** $p(x|-5) = \frac{2}{5}(5 - x) \qquad 0 < x \le 5$
$$= 0 \qquad\qquad\quad \text{elsewhere}$$
$$p(x|0) = \dfrac{10 - x}{50} \qquad 0 < x \le 10$$
$$= 0 \qquad\qquad\quad \text{elsewhere}$$
$$p(x|5) = \tfrac{1}{75}(5 + x) \quad 0 < x \le 5$$
$$= \tfrac{1}{75}(15 - x) \quad 5 < x \le 10$$
$$= 0 \qquad\qquad\quad \text{elsewhere}$$

   b. $x = 0$

3-7   **a.** 85   **b.** 37   **c.** $\sigma^2_{X+Y} = 37 \quad \sigma^2_{X-Y} = 85$

3-8 $-.269$

3-10 **a.** $p_\Phi(\phi) = \dfrac{\phi}{4\pi} \qquad 0 < \phi \le 2\pi$

$$= \dfrac{4\pi - \phi}{4\pi^2} \quad 2\pi < \phi \le 4\pi$$
$$= 0 \qquad\quad \text{elsewhere}$$

   b. $p_\Phi(\phi) = \dfrac{1}{2\pi} \qquad 0 < \phi \le 2\pi$
$$0 \qquad\quad \text{elsewhere}$$

3-11 **a.** $e^{-u^2\sigma^2/2}$

Chapter 4

4-1   **a.** $\frac{1}{8}$ **b.** 0 and E; no

4-2   **a.** over one hour period—continuous, nondeterministic, stationary, ergodic; over a ten year period—continuous, nondeterministic, nonstationary, nonergodic   **b.** over a 10 minute period—discrete, nondeterministic, stationary, ergodic; over a 24 hour period—discrete, nondeterministic, nonstationary, nonergodic; over a 360 day period—discrete, nondeterministic, nonstationary, nonergodic   **c.** continuous, deterministic, nonstationary, nonergodic   **d.** continuous, deterministic, nonstationary, nonergodic   **e.** continuous, deterministic, stationary, nonergodic

4-3   **a.** $P(x) = \dfrac{3}{4} + \dfrac{x}{4A} \quad 0 < x \le A$

b. $p(x) = \dfrac{3}{4}\,\delta(x) + \dfrac{1}{4A}\,[u(x) - u(x - A)]$

c. $\bar{X} = A/8;\ \overline{X^2} = A^2/12;\ \sigma^2 = 13A^2/192$ **d.** $\langle x \rangle = A/8;\ \langle x^2 \rangle = A^2/12$
e. yes, yes

4-4   **a.** discrete, deterministic, stationary, ergodic   **b.** $p_X(x) = \dfrac{7}{8}\,\delta(x) +$

$\dfrac{1}{8}\,\delta(x - y)$ **c.** $\bar{X} = 0,\ \overline{X^2} = \tfrac{1}{8}$ **d.** $\langle x \rangle = Y/8,\ \langle x^2 \rangle = Y^2/8$

e. Process is stationary but not ergodic.

4-5   **a.** 200.3   **b.** 200   **c.** 4.9   **d.** 69.34
4-6   **a.** $184 < \bar{X} < 216$ **b.** 8,000

Chapter 5

5-1   **a.** $R_X(\tau) = \dfrac{A^2}{t_a}\,[b - |\tau|]\quad |\tau| \le b$

$= 0 \qquad\qquad\quad |\tau| > b$
b. Same as part a.   **c.** Yes. $R_X(\tau) = \mathcal{R}_x(\tau)$ **d.** $A^2 b/t_a$

5-2   **a.** $R_V(\tau) = \dfrac{10^4}{3}\left[\dfrac{1}{2} - \dfrac{|\tau - NT|}{T}\right]\quad |\tau - NT| \le \dfrac{T}{2}\quad N = 0, \pm1, \pm2, \dots$

b. $\mathcal{R}_v(\tau) = X^2\left[\dfrac{1}{2} - \dfrac{|\tau - NT|}{T}\right]\quad |\tau - NT| \le \dfrac{T}{2}\quad N = 0, \pm1, \pm2, \dots$

5-3 $\dfrac{10^8}{9}\left[\dfrac{1}{2} - \dfrac{|\tau - NT|}{T}\right]\quad |\tau - NT| \le \dfrac{T}{2}\quad N = 0, \pm1, \pm2, \dots$

5-4   **a.** $\overline{X^2} = 5,\ \sigma^2 = 5$ **b.** $f_0 = \tfrac{3}{2}$ **c.** They are positively correlated.
d. $\cong .415$ sec.   **e.** $\tfrac{1}{4}$

5-5   **a.** $\hat{r} = \dfrac{R_{XY}(\tau)}{\sigma_Y{}^2}$ **b.** $|\hat{r}| \le \dfrac{\sigma_X}{\sigma_Y}$

5-6   None of them.

5-7   **a.** Process is stationary but nonergodic.   **b.** $4\,\dfrac{\sin 5\tau}{5\tau}$

5-8   **a.** $13(9 + e^{-3\tau^2})(2e^{-2|\tau|}\cos\omega\tau)$ **b.** $\bar{Z} = 0,\ \sigma_Z{}^2 = 260$
5-9   **a.** $R_{XY}(\tau) = \tfrac{1}{2}AB\sin\omega_0\tau;\ R_{YX}(\tau) = -\tfrac{1}{2}AB\sin\omega_0\tau$ **b.** $X(t)$ and $Y(t)$ are
orthogonal.

5-10   **a.** $R_Y(\tau) = \dfrac{A^2}{2}\left[\dfrac{b - |\tau|}{t_a}\right] \qquad\qquad |\tau| < b$

$= \dfrac{A^2}{4}\left[\dfrac{b - |\tau - Nt_a|}{t_a}\right]\quad |\tau - Nt_a| < b\quad N = \pm1, \pm2, \dots$

$= 0 \qquad\qquad\qquad\qquad\text{elsewhere}$

b. $R_{XY}(\tau) = \dfrac{A^2}{2}\left[\dfrac{b - |\tau|}{t_a}\right]\quad |\tau| \le b$

$= 0 \qquad\qquad\quad |\tau| > b$
c. $R_{YX}(\tau) = R_{XY}(\tau)$
5-11 $R_{Z_1}(\tau) = R_X(\tau) + R_Y(\tau);\ R_{Z_1Z_2}(\tau) = R_X(\tau) - R_Y(\tau);\ R_{Z_1Z_2}(\tau) = R_X(\tau) -$
$R_Y(\tau);\ R_{Z_2}(\tau) = R_X(\tau) + R_Y(\tau)$

5-12 **a.** $\Lambda = \begin{bmatrix} 1 & 0 & -e^{-1} \\ 0 & 1 & 0 \\ -e^{-1} & 0 & 1 \end{bmatrix}$

b. $\Lambda = \begin{bmatrix} 1 & e^{-.5}\cos .5\pi & e^{-1}\cos\pi & \cdots & e^{-.5(N-1)}\cos[.5(N-1)\pi] \\ e^{-.5}\cos .5\pi & & & & \\ \cdot & & & & \\ \cdot & & & & \\ \cdot & & & & \\ e^{-.5(N-1)}\cos[.5(N-1)t] & \cdots & & & 1 \end{bmatrix}$

$(\Lambda)_{ij} = e^{-.5|i-j|}\cos[.5|i-j|\pi]$

Chapter 6

6-1 **a.** Not valid, can be negative **b.** Valid **c.** Not valid, not even **d.** Not valid, not real **e.** Valid **f.** Valid

6-2 $S_X(\omega) = \dfrac{\pi}{2}\left[a_0{}^2\delta(\omega) + \displaystyle\sum_{n=1}^{\infty} \{(a_n{}^2 + b_n{}^2)[\delta(\omega - n\omega_0) + \delta(\omega + n\omega_0)]\}\right]$

6-3 **a.** $S_X(\omega) = \dfrac{A^2T}{2}\left[\dfrac{\sin\dfrac{\omega T}{4}}{\dfrac{\omega T}{4}}\right]^2$

b. $S_X(\omega) = \left(\dfrac{\sin\omega T/4}{\omega T/4}\right)^2\left[\dfrac{3A^2T}{2} + 2\pi A^2 \displaystyle\sum_{n=-\infty}^{\infty}\delta\left(\omega - \dfrac{4\pi n}{T}\right)\right]$

6-4 $S_X(\omega) = \dfrac{t_1{}^2}{T}\left(\dfrac{\sin\omega t_1/2}{\omega t_1/2}\right)^2$

6-5 **b.** $\dfrac{-s^2 + 10}{s^4 - 5s^2 + 4}$ **e.** $\left[\dfrac{\sinh s}{s}\right]^2$ **f.** $\delta(s) + \dfrac{-s^2}{s^4 + 1}$

6-6 **b.** 1 **e.** 0.5 **f.** $\dfrac{1}{2\pi} + \dfrac{\sqrt{2}}{4}$

6-7 $\frac{1}{6}$

6-8 $S_X(\omega) = \dfrac{A\alpha}{(\omega - \omega_0)^2 + \alpha^2} + \dfrac{A\alpha}{(\omega + \omega_0)^2 + \alpha^2}$

6-9 $S_X(\omega) = \dfrac{T}{16}\left[\dfrac{\sin\omega T/8}{\omega T/8}\right]^2$

6-11 **a.** $S_Z(\omega) = S_X(\omega) + S_{XY}(\omega) + S_{YX}(\omega) + S_Y(\omega)$ **b.** $S_{XY}(\omega) = S_{YX}(\omega) = 2\pi\bar{X}\bar{Y}\delta(\omega)$ **c.** $S_{XZ}(\omega) = S_X(\omega) + S_{XY}(\omega)$

6-12 **a.** $0.5_r\hat{S}_X(\omega) + 0.25\left[{}_r\hat{S}_X\left(\omega + \dfrac{\pi}{\tau_m}\right) + {}_r\hat{S}_X\left(\omega - \dfrac{\pi}{\tau_m}\right)\right]$ **b.** The Hamming-window's highest sidelobe is approximately one-third of the height of the Hanning-window's highest sidelobe.

6-13 $\omega_1 \simeq \dfrac{12.56}{t_1}$

6-15 $R_X(\tau) = 2\,WS_0\,\dfrac{\sin 2\pi\,W\tau}{2\pi W\tau}$

$R_X\left(\dfrac{k}{2W}\right) = \begin{cases} 2WS_0 & k = 0 \\ 0 & k \neq 0 \end{cases}$

$\Lambda = 2WS_0\,\mathbf{I}_n$; $\mathbf{I}_n$ is the identity matrix of order n.

Chapter 7

7-1   **a.** $h(t) = \frac{9}{4}[e^{-2t} - e^{-18t}]u(t)$ **b.** 1   **c.** $\dfrac{9S_0}{10}$

7-2   **a.** $h(t) = u(t) - u(t - T)$ **b.** T **c.** yes   **d.** $S_0 T$

7-3 $\bar{Y} = T/2;\ \sigma_Y{}^2 = T^2/12 + \dfrac{1}{2\pi^2}\sin^2 \pi T$

7-4 $R_Y(\tau) = S_0\left[\delta(\tau) - \dfrac{R}{2L}\,e^{-\frac{R}{L}|\tau|}\right]$

7-5 $R_Y(\tau) = \dfrac{\beta^2 S_0}{2\left[\left(\dfrac{R}{L}\right)^2 - \beta^2\right]}\left[\dfrac{R}{L}\,e^{-\frac{R}{L}|\tau|} - \beta e^{-\beta|\tau|}\right]$

7-6 $R_Y(\tau) = \frac{1}{3} + \dfrac{b^2\cos 2\pi\tau}{2(b^2 + 4\pi^2)};\ b = \dfrac{1}{RC}$

7-7 $R_{XY}(\tau) = S_0\left[\delta(\tau) - \dfrac{R}{L}\,e^{-\frac{R}{L}\tau}\,u(\tau)\right]$

$R_{YX}(\tau) = S_0\left[\delta(-\tau) - \dfrac{R}{L}\,e^{\frac{R}{L}\tau}\,u(-\tau)\right]$

7-8 $R_{XY}(\tau) = \dfrac{\beta S_0}{2(\beta^2 - b^2)}\left\{(\beta^2 + b\beta)e^{-\beta\tau} - 2b\beta e^{-b\tau}\right\}$ $\tau \geq 0$

$= \dfrac{\beta^2 S_0}{2(b + \beta)}\,e^{\beta\tau}$ $\tau < 0$

$R_{YX}(\tau) = \dfrac{\beta^2 S_0}{2(b + \beta)}\,e^{-\beta\tau}$ $\tau \geq 0$

$= \dfrac{\beta S_0}{2(\beta^2 - b^2)}\left\{(\beta^2 + b\beta)e^{+\beta\tau} - 2b\beta e^{+b\tau}\right\}$ $\tau < 0$

7-9 $R_{YZ}(\tau) = \dfrac{S_0 b}{2}\,e^{-b\tau}$ $\tau \geq 0$

$= \dfrac{-S_0 b}{2}\,e^{+b\tau}$ $\tau < 0$

7-10 $\frac{10}{9}$

7-11   **a.** $h(t) = \frac{9}{4}[e^{-2t} - e^{-18t}]\,u(t)$ **b.** 1   **c.** $9S_0/10$

7-12 $S_0 T$

7-13 $S_Y(\omega) = S_0 - \dfrac{b^2 S_0}{\omega^2 + b^2}$

$R_Y(\tau) = S_0 \delta(\tau) - \dfrac{S_0 b}{2} e^{-b|\tau|}$

7-14 $S_X(\omega) = 10^{-5}/\pi$

7-15 $R_Y(\tau) = 2N_0 W \cos 2\pi f_0 \tau \dfrac{\sin \pi W \tau}{\pi W \tau}$

7-17 $B = \dfrac{W}{2} \sqrt{\pi/\ln 2} = \dfrac{\sigma}{4\sqrt{\pi}}$

7-18 **a.** kT/C **b.** The result does not depend on R, because as R changes, the dependence of the noise spectral density on R is cancelled by the dependence of the circuit bandwidth on R.

7-19 **b.** 1.245×10^5 volts **c.** 55.7×10^{-6} volts

Chapter 8

8-1 **a.** B **b.** A **c.** B **d.** B **e.** A **f.** A **g.** A **h.** B **i.** B **j.** B

8-3 $b = \omega_0$

8-4 $b = 2$

8-5 **a.** $h(t) = s(1 - t)u(t)$ **b.** 150 **c.** $h(t) = s(-t)u(t)$; 60

8-6 **a.** 50 **b.** $t_0 \simeq 2$

8-7 $\dfrac{.485s + .578}{(s + \sqrt{3})(s + 2\sqrt{3})}$

Index

Index

A

Analog-to-digital conversion, errors in, 55
A posteriori probability, continuous, 80
  discrete, 23
Applications of probability, 1–5
A priori probability, continuous, 80
  discrete, 23
Autocorrelation function, 108
  binary process, 110
  of derivatives, 121
  examples of, 117
  measurement of, 115
  properties of, 112
  relation to spectral density, 146, 233
  of system output, 172
Automatic control system, 189
Axioms, of probability, 17

B

Band-limited white noise, 151
Bandwidth, equivalent-noise, 187
  half-power, 191
Bayes' Theorem, continuous, 79
  discrete, 23
Bernoulli distribution, 226
Bernoulli trials, 27
Beta distribution, 227

Binary signals, autocorrelation
    function, 110
  probability density function, 59
  spectral density, 159
Binomial coefficient, 28
  table of, 231
Binomial distribution, 226

C

Cartesian product space, 26
Cauchy distribution, 227
Central limit theorem, 48
Central moments, 44
Certain event, 9, 17
Characteristic function, 88
  definition, 89
  moments from, 90
  of sums, 89
  for two random variables, 91
Chi-square distribution, 53, 228
Combined experiments, 25
Complement, of sets, 15
Conditional expectation, 63
  mean, 63
Conditional probability, 10, 19
  density and distribution functions, 60
  for two random variables, 78

Continuous probability, 7
Continuous random processes, 96
Contour integration, 234
Convolution, of density functions, 86
Correlation, 77, 83
  coefficient, 83
  for independent random variables, 82
  matrix, 124
Correlation functions, 107
  autocorrelation, 108
  cross-correlation, 108, 119
  of derivatives, 121
  in matrix form, 124
  table of spectral density pairs, 233
  time autocorrelation, 109
Covariance, 83
  matrix, 124
Cross-correlation functions, 119
  of derivatives, 121
  examples of, 122
  between input and output of a system, 176
  properties of, 120
Cross-spectral density, 152

D

Definite integrals, table of, 223
Delta distributions, 58
  binary, 59
  multi-level, 59
DeMoivre-Laplace theorem, 30
DeMorgan's laws, 15
Density function, probability, 39
  definition of, 40
  properties of, 40
  table of, 226–230
  transformations, 41
  types of, 40
Deterministic random processes, 97
Discrete probability, 7
Discrete random processes, 96
Distribution function, probability, 37
  definition of, 37
  properties of, 38
  table of, 226–230
  types of, 38

E

Electronic voltmeter, scale factor, 73
Elementary event, 17
Elements, of a set, 12
Empty set, 13
Ensemble, 36, 95
  averaging, 43
Equivalent-noise bandwidth, 187
Ergodic random processes, 100
Erlang distribution, 58, 228
Erlang random variable, 58
Error function, 80
Estimates, autocorrelation function, 115
  mean value, 101
  signal in noise, 79
  spectral density, 153
  variance, 103
Estimates, mean of, 102, 116, 157
  variance of, 102, 103, 116, 158
Estimation, from $p(x|y)$, 80
Event, certain, 9, 17
  definition of, 7
  elementary, 17
  impossible, 9, 17
  probability of, 17
  for a random variable, 37
Expected value, 43
Experiment, 6
Exponential distribution, 56, 228

F

Finite-time integrator, 179
Fourier transform, 131
  relation to spectral density, 133
  table of, 224
Frequency-domain analysis of systems, 183–187
  cross-spectral density, 187
  examples of, 187–192
  mean-square value, 186
  spectral density, 183

G

Gamma distribution, 58, 228
Gaussian random process, 103
  multi-dimensional, 126, 229

Gaussian random variable, 46
  density function, 46
  distribution function, 47
  moments of, 48
General moments, 44

H

Hamming window function, 156

I

Identities, trigonometric, table of, 221
Impossible event, 9, 17
Impulse response, 167
  system output, 168
Indefinite integrals, table of, 222
Independence, statistical, 11, 24
  more than two events, 24
  of random variables, 82
Integrals, definite, table of, 223
Integrals, indefinite, table of, 222
Integration, in the complex plane, 234
Interchange of integrals, 169
Intersection of sets, 14

J

Joint characteristic function, 91
Joint probability, 9
  density function, 75
  distribution function, 75
  example of, 76
  expected values from, 76
  multi-dimensional Gaussian, 126
Jordan's lemma, 238

L

Lag window, 154
  Hamming, 156
  Hanning, 165
  rectangular, 155
Laplace distribution, 228
Laplace transform, table of, 225

Laurent expansion, 235
Log-normal distribution, 53, 229

M

Marginal probability, 10
  density function, 77
Matched filters, 207
  exponential pulse, 208
  periodic pulses, 209
  rectangular pulse, 207
Mathematical tables, 221–225
  binomial coefficients, 231
  correlation function—spectral
    density pairs, 233
  definite integrals, 223
  Fourier transforms, 224
  indefinite integrals, 222
  Laplace transforms, one-sided, 225
  normal probability distribution, 234
  trigonometric identities, 221
Matrix correlation, 124
  covariance, 125
Maxwell distribution, 52, 229
Mean-square value, 44
  from residues, 145
  from spectral density, 142
  of system output, 169
  table for, 143
Mean value, 43
  from autocorrelation function, 113
  conditional, 63
  from spectral density, 138
  of system output, 169
Measurement of spectral density, 153–158
  discrete approximations, 157
  use of window functions, 155, 156
  variance of estimate, 158
Moments, 44
Mutually exclusive events, 8
  sets, 15

N

Noise figure, 195
Noise, thermal agitation, 194
Nondeterministic random processes, 97

Nonergodic random processes, 100
Nonstationary random processes, 98
Normal probability distribution, 46, 229
  bivariate, 229
  table of, 232
Normalized covariance, 83
Null set, 13

O

Optimum linear systems, 198–215
  criteria for, 197
  maximum signal-to-noise ratio, 205
  minimum mean-square error, 210
  parameter adjustment, 199
  restrictions on, 199
Outcome, of an experiment, 7

P

Parseval's theorem, 133
Pascal distribution, 227
Poisson distribution, 227
Power distribution, 49
Probability, applications of, 1–5
  a posteriori, 23
  a priori, 23
  axiomatic, 17–19
  conditional, 10, 19
  continuous, 7
  definitions of, 5, 6
  discrete, 7
  joint, 9
  marginal, 9
  relative-frequency, 6–12
  space, 17
  total, 21
Probability density function, 39
  definition of, 40
  marginal, 77
  properties of, 40
  of the sum of random variables, 85
  table of, 226–230
  transformations, 41
  types of, 40
Probability distribution function, 37
  definition of, 37
  marginal, 39

Probability distribution function
  (*Continued*)
  properties of, 38
  table of, 226–230
  types of, 38
Probability distributions, table of, 226–230
  Bernoulli, 226
  beta, 227
  binomial, 226
  Cauchy, 227
  chi-square, 228
  Erlang, 228
  exponential, 228
  gamma, 228
  Laplace, 228
  log-normal, 229
  Maxwell, 229
  normal, 229
  normal bivariate, 229
  Pascal, 227
  Poisson, 227
  Rayleigh, 229
  uniform, 230
  Weibull, 230
Process, random, 36, 95
  measurement of, 100
  types of, 96

Q

Quantizing error, 55

R

Radar, pulse amplitudes, 72
  cross-correlation, 123
Raised-cosine signals, 159
  spectral density, 160
Random process, 36, 95
  measurement of, 100
  types of, 96
Random variable, concept of, 35
  Gaussian, 46
Rational spectral density, 136
Rayleigh distribution, 51, 229
  application to radar, 72
  application to traffic, 68
Rectangular window function, 155
Relative-frequency approach, 6–12

Reproducible density functions, 88
Residue, 146, 236

S

Sample function, 95
Sampling theorem, 152
Schwarz inequality, 206
Set theory, 12
Sets, complement of, 15
   difference of, 15
   disjoint, 15
   empty, 13
   equality of, 13
   intersection of, 14
   mutually exclusive, 15
   null, 13
   product of, 14
   sum of, 13
   union of, 13
   Venn diagram, 13
Simpson density function, 71
Singular point, 235
   isolated, 235
Space, of a set, 12
Spectral density, 131
   applications of, 158–163
   in the complex frequency plane, 141
   for a constant, 138
   cross-spectral density, 152
   definition of, 134
   for a derivative, 140
   mean-square value from, 142
   measurement of, 153–158
   for a periodic process, 138
   properties of, 136
   for a pulse sequence, 139
   rational, 136
   relation to the autocorrelation function,
     146
   white noise, 136
Standard deviation, 44
Standardized variable, 84
Stationary random processes, 98
   in the wide sense, 99
Statistical independence, 11, 24
   more than two events, 24
   of random variables, 82

Subset, 12
   as events, 17
Sums of random variables, mean of, 85
   density function of, 86
   Gaussian, 87
   variance of, 85
System analysis, 167–192
   frequency-domain methods, 183–187
   time-domain methods, 168–179

T

Thermal agitation noise, 194
Time correlation functions, 109
   cross-correlation, 120
Time-domain analysis of systems, 168–179
   autocorrelation function, 172
   cross-correlation function, 176
   examples of, 179–182
   mean value, 169
   mean-square value, 170
Total probability, 21
Traffic measurement, 68
Transfer function, of a system, 167
Transformation of variables, 41
   logarithm, 53
   square-law, 42
Trial, 7
Triangular density function, 71
Trigonometric identities, table of, 221
Two random variables, 74

U

Unbiased estimate, 116
Uniform distribution, 55, 230
Union of sets, 13

V

Variables, random, 35
Variance, 44
   estimate of, 103
Venn diagram, 13

Voltmeter resistor, 66
Voltmeters, electronic, 73

W

Weibull distribution, 230
White noise, 136, 150
  approximation to, 175
  autocorrelation function of, 150

White noise (*Continued*)
  band-limited, 151
  uses of, 150
Wiener filter, 213
  error from, 214
Wiener-Khinchine relation, 148

Z

Zener diode circuit, 64

NOTES

NOTES

NOTES

NOTES

NOTES

NOTES

NOTES

NOTES

NOTES

NOTES

NOTES

NOTES

NOTES

NOTES

NOTES

NOTES